The New Carboı

Antipode Book Series

General Editor: Dr Rachel Pain, Reader in the Department of Geography, Durham University, UK

Like its parent journal, the Antipode Book Series reflects distinctive new developments in radical geography. It publishes books in a variety of formats – from reference books to works of broad explication to titles that develop and extend the scholarly research base – but the commitment is always the same: to contribute to the praxis of a new and more just society.

Published

The New Carbon Economy: Constitution, Governance and Contestation
Edited by Peter Newell, Max Boykoff and Emily Boyd

Capitalism and Conservation
Edited by Dan Brockington and Rosaleen Duffy

Spaces of Environmental Justice
Edited by Ryan Holifield, Michael Porter and Gordon Walker

The Point is to Change It: Geographies of Hope and Survival in an Age of Crisis
Edited by Noel Castree, Paul Chatterton, Nik Heynen, Wendy Larner and Melissa W. Wright

Privatization: Property and the Remaking of Nature-Society
Edited by Becky Mansfield

Practising Public Scholarship: Experiences and Possibilities Beyond the Academy
Edited by Katharyne Mitchell

Grounding Globalization: Labour in the Age of Insecurity
Edward Webster, Rob Lambert and Andries Bezuidenhout

Privatization: Property and the Remaking of Nature-Society Relations
Edited by Becky Mansfield

Decolonizing Development: Colonial Power and the Maya
Joel Wainwright

Cities of Whiteness
Wendy S. Shaw

Neoliberalization: States, Networks, Peoples
Edited by Kim England and Kevin Ward

The Dirty Work of Neoliberalism: Cleaners in the Global Economy
Edited by Luis L. M. Aguiar and Andrew Herod

David Harvey: A Critical Reader
Edited by Noel Castree and Derek Gregory

Working the Spaces of Neoliberalism: Activism, Professionalisation and Incorporation
Edited by Nina Laurie and Liz Bondi

Threads of Labour: Garment Industry Supply Chains from the Workers' Perspective
Edited by Angela Hale and Jane Wills

Life's Work: Geographies of Social Reproduction
Edited by Katharyne Mitchell, Sallie A. Marston and Cindi Katz

Redundant Masculinities? Employment Change and White Working Class Youth
Linda McDowell

Spaces of Neoliberalism
Edited by Neil Brenner and Nik Theodore

Space, Place and the New Labour Internationalism
Edited by Peter Waterman and Jane Wills

Forthcoming

Banking Across Boundaries: Placing Finance in Capitalism
Brett Christophers

Fat Bodies, Fat Spaces: Critical Geographies of Obesity
Rachel Colls and Bethan Evans

Gramscian Geographies: Space, Ecology, Politics
Edited by Michael Ekers, Gillian Hart, Stefan Kipfer and Alex Loftus

Places of Possibility: Property, Nature and Community Land Ownership
Fiona D. Mackenzie

Radical Democratization: Inventing Networks of Equivalence
Mark Purcell

The New Carbon Economy

Edited by

Peter Newell, Max Boykoff and Emily Boyd

WILEY-BLACKWELL

A John Wiley & Sons, Ltd., Publication

This edition first published 2012
Originally published as Volume 43, Issue 3 of *Antipode*
Chapters © 2012 The Authors
Book compilation © 2012 Editorial Board of Antipode and Blackwell Publishing Ltd

Blackwell Publishing was acquired by John Wiley & Sons in February 2007. Blackwell's publishing program has been merged with Wiley's global Scientific, Technical, and Medical business to form Wiley-Blackwell.

Registered Office
John Wiley & Sons Ltd, The Atrium, Southern Gate, Chichester, West Sussex, PO19 8SQ, United Kingdom

Editorial Offices
350 Main Street, Malden, MA 02148-5020, USA
9600 Garsington Road, Oxford, OX4 2DQ, UK
The Atrium, Southern Gate, Chichester, West Sussex, PO19 8SQ, UK

For details of our global editorial offices, for customer services, and for information about how to apply for permission to reuse the copyright material in this book please see our website at www.wiley.com/wiley-blackwell.

The right of Peter Newell, Max Boykoff and Emily Boyd to be identified as the authors of the editorial material in this work has been asserted in accordance with the Copyright, Designs and Patents Act 1988.

Wiley also publishes its books in a variety of electronic formats. Some content that appears in print may not be available in electronic books.

Designations used by companies to distinguish their products are often claimed as trademarks. All brand names and product names used in this book are trade names, service marks, trademarks or registered trademarks of their respective owners. The publisher is not associated with any product or vendor mentioned in this book. This publication is designed to provide accurate and authoritative information in regard to the subject matter covered. It is sold on the understanding that the publisher is not engaged in rendering professional services. If professional advice or other expert assistance is required, the services of a competent professional should be sought.

Library of Congress Cataloging-in-Publication data is available for this book.

The new carbon economy / edited by Peter Newell, Max Boykoff and Emily Boyd.
    p.   cm.
  Includes index.
  "Originally published as Volume 43, Issue 3 of Antipode."
  ISBN 978-1-4443-5022-7 (pbk.)
1. Energy industries–Environmental aspects.   2. Energy consumption–Environmental aspects.
3. Energy policy–Environmental aspects.   4. Carbon offsetting.   5. Carbon dioxide mitigation.
6. Environmental policy.   7. Climactic changes–Economic aspects.   I. Newell, Peter (Peter John)
II. Boykoff, Maxwell T.   III. Boyd, Emily.   IV. Antipode.
  HD9502.A2N485 2012
  363.738'746–dc23
                                                                                    2011038340

Set in 11pt Times by Aptara Inc., New Delhi, India.
Printed in Malaysia by Ho Printing (M) Sdn Bhd

1   2012

Contents

List of Contributors

Ian Bailey	University of Plymouth
Emily Boyd	University of Reading
Maxwell Boykoff	University of Colorado
Adam G. Bumpus	University of Melbourne
Philippe Descheneau	University of Ottawa
Andy Gouldson	University of Leeds
María Gutiérrez	Consultant
David M. Lansing	University of Maryland
Heather Lovell	University of Edinburgh
Donald MacKenzie	University of Edinburgh
Peter Newell	University of Sussex
Matthew Paterson	University of Ottawa
Roopali Phadke	Macalester College

Chapter 1
The "New" Carbon Economy: What's New?

Emily Boyd, Maxwell Boykoff and Peter Newell

Introduction

We now have what is commonly called a carbon economy. However, it is in fact made up of several, increasingly inter-connected, carbon markets. It takes different forms in different parts of the world, but includes systems of emissions trading (in the EU, some states in the USA and emerging schemes in cities, such as Montreal), and the buying and selling of offsets through United Nations-controlled "compliance" markets, most notably though the Clean Development Mechanism (CDM) created by the Kyoto Protocol, as well as through "voluntary" markets. The carbon economy has had a turbulent history: its monetary value was affected by global financial meltdown, which also suppressed levels of demand for carbon credits, and its legitimacy questioned amid claims of climate fraud, "toxic carbon", and acts of (neo)colonial dispossession (Bachram 2004; Friends of the Earth 2009; Lohmann 2005, 2006).

And yet the importance of the carbon economy should not be underestimated. With the CDM, for example, Certified Emissions Reductions (CERs) amounting to more than 2.7 billion tonnes of carbon dioxide equivalent are expected to be produced in the first commitment period of the Kyoto Protocol (2008–2012) (United Nations Framework Convention on Climate Change—UNFCCC 2010). The revenues of the CDM constitute the largest source of mitigation finance to developing countries to date (World Bank 2010). Over the 2001–2012 period, CDM projects could raise US$15—24 billion in direct carbon revenues for

The New Carbon Economy, First Edition. Edited by Peter Newell, Max Boykoff and Emily Boyd.

developing countries. Actual revenues will, of course, depend on the price of carbon. The voluntary carbon market, meanwhile, saw early exponential growth with a tripling of transactions between 2006 and 2007, when it was worth US$331 million. It remains only fraction of the size of regulated markets though, and had its value nearly halve in 2009 to US$387 million and fall by a quarter in volumes of carbon transacted (Capoor and Ambrosi 2009). These markets remain important but very unstable. What is perhaps most notable is that, despite these crises, faith in carbon markets as a key element of global responses to the threat of climate change remains strong, as affirmed by the UN climate change meeting in December 2010 in Cancún.

The world of climate politics was not always thus. In the years up to 1992, being built on foundations similar to those of other multilateral environmental agreements, and particularly coming in the wake of the apparently successful ozone regime, a command and control process of setting targets globally that countries enforce nationally seemed a logical way to proceed for the climate regime. This after all was the template for numerous previous regimes aimed at regulating pollutants of one form or another. Yet, even as discussions moved from the UNFCCC to efforts to produce a legally binding emissions reductions treaty, opposition set in. This was not an issue amenable to convenient techno-fixes such as the substitution of damaging chemicals (replacing CFCs with HCFCs as in the case of ozone) or one that only affected a handful of large multinational enterprises in the core of the global economy for whom alternative accumulation strategies could easily be identified. The ramifications of regulating energy supply and use to the world's economy, upon which growth depends, made climate change a "wicked" policy challenge.

It soon became clear that if action were to be taken at all, and particularly by those economies contributing most to the problem, markets offered the most politically acceptable solution as part of a suite of measures to bring down the costs associated with reducing emissions and increase flexibility about where emissions' reductions take place. Market-based solutions aligned closely with prevailing ideologies regarding regulation and the primacy of "efficiency" and the preferences of powerful fractions of capital in leading economies for trading schemes over taxes or regulation. Their incorporation became the *quid pro quo* for the involvement of the USA in particular (notwithstanding the fact that having insisted on flexible mechanisms, the USA then walked away from the Kyoto deal). The USA fought hard to project the "success" of its sulphur dioxide trading scheme as a model for carbon trading, while domestic industries sought to avoid domestic reduction measures that might drive capital overseas and had a clear preference for paying others to reduce where it is cheapest to do so (Gilbertson and Reyes 2009). The prospect of a transfer of resources from the global North to South through trading of offsets also provided an incentive for many developing countries' governments to support market-based approaches. Emissions trading and the CDM were what eventually resulted from this confluence of potential beneficiaries from the carbon economy.

Businesses, meanwhile, came under pressure to demonstrate their contributions to tackling climate change. After many years of hostile opposition to action (Newell and Paterson 1998), some began to see a business case for action on climate change aimed at meeting demand for lower carbon products and services as well as providing carbon offset projects for clients and consumers wanting to claim "carbon neutrality". Voluntary carbon offsets met this need.

Told this way, it would appear that the "new" carbon economy is a relatively novel phenomenon, though offset providers will remind you that the first carbon offset project dates as far back as 1989. Many claims to novelty are made about the "new" carbon economy nevertheless, not least by the actors and beneficiaries of the economy themselves who seek to identify new opportunities in the reduction of carbon and technological innovation. It is also the case that the political, scalar and temporal challenges of climate change are often deemed to be novel and unprecedented (Giddens 2009). For some, the "new" carbon economy is capable of overcoming resistance to action on climate change by generating new sites of accumulation or by linking "local" and "global" spaces of climate governance, through offsets, for example.

What we have briefly described above is illustrative of an important transition that has taken place in the nature of responses to climate change that reflects and advances a particular form of neo-liberalism. The carbon economy may be novel in the sense that it would not have been politically viable or imaginable even 20 years earlier, and because it bears the hallmarks of a specific period of capitalist development in terms of the confluence of forces in engages (finance capital in particular) and the characteristics it displays (reliance on networks, globalised market opportunities) (Newell and Paterson 2009). While there are certainly precedents for the creation of markets to address resource overconsumption (around fisheries and in the area of conservation for example), in terms of its reach and scale (geographically and sectorally), the range of powerful actors now engaged and the levels of finance being circulated in the carbon economy, claims of newness go beyond describing the banal existence of a newly created market and describe instead the emergence of an historically unparalleled experiment in marketised environmental governance. Many of the politics and actors which created and sustain the carbon economy and the political and social challenges they create, may be less novel, however. So what is really new about the "new" carbon economy?

There remains a clear need for researchers to subject claims of *newness* and *uniqueness* to critical scrutiny. By this we refer to the need to historicise such claims, identify precedents, explore them comparatively and reveal embedded assumptions and politics. This is the purpose of this book. We suggest three areas in particular where claims of novelty arise and which require further attention: the ecological and social inputs which produce, sustain and constitute the new carbon economy; the governance issues and challenges that arise from the commodification and trading of carbon; and the effects it generates: the winners and losers from this economy. We can summarise this trio as novelty in the *constitution*, *governance* and *effects* of the new carbon economy.

Constituting the "New" Carbon Economy

Many of the contributions to this book focus on the materiality of carbon (such as that by Bumpus): its physical properties and how this affects its enrolment into the global circuits of capital. Whether emphasising its "uncooperative nature" (Bakker 2004), how difficult it is to measure or to control in ways that commodification demands, or the failed articulations between global capital and local socio-ecological systems that Lansing explores, many contributions highlight the need to take seriously the biophysical properties of carbon. These properties may give rise to new governance challenges (in terms of commodification, measurement and exchange) or produce distinct forms of socio-ecological relations. So what's the "matter" with carbon?

Focusing on carbon can be seen as somewhat reductionist: there are greenhouse gases that do not contain carbon (e.g. nitrous oxide), and not all carbon-containing emissions (e.g. carbon monoxide) trap heat. However, markets in greenhouse gas reductions are organised around carbon dioxide equivalence to create "exchange value" and a fungible commodity that can be traded across products and projects. As Gavin Bridge has noted, carbon has become:

> a common denominator for thinking about the organization of social
> life in relation to the environment ... from fossil-fuel addiction and
> peak oil to blood barrels and climate change, carbon's emergence as
> a dominant optic for thinking and writing about the world and human
> relations within it is tied to the various emergencies with which it is
> associated (Bridge 2010:2).

Carbon is literally embedded in different territories, histories, economies and politics of which climate change is merely a latest addition. It is part of a cycle involving forests and oceans: it moves and is not static, part of what Lansing refers to in this book as the "precarious choreography of the global carbon cycle". These properties matter. This leads to framing and boundary contests over, for example, what count as forests (Gutiérrez, this volume), the extent to which forests hold carbon, for how long and on whose behalf and how responses to climate change physically and politically reconfigure landscapes (Phadke, this volume). Decisions to preserve carbon stocks in forests and land, or to keep it in the ground, imply huge social and economic trade-offs for those whose livelihoods depend on carbon in fossil fuels. Moreover, a great deal of work—politically and literally in the form of labour—goes into producing, extracting value from, and commodifying, materials which contain carbon, as well products which claim to prevent the release of carbon; offsets that can be bought and sold. Attempts to regulate carbon often pitch industries such as coal alongside trade unions, such as those representing mineworkers, against environmentalists calling for limits on use. Workers may be well aware that their labour produces surplus value for their employees from which they are excluded, but the fate of the industry represents their own fate. These histories, conflicts and social dynamics form the basis of the forms of governance that have been created to manage the carbon economy and to keep carbon capital in circulation.

Governing the "New" Carbon Economy

Many claims of newness in relation to the new carbon economy relate to its governance dimensions. Critics focus on its lack of governance, its unregulated and "wild west" nature. This is considered particularly problematic when the system is premised on claims of additionality: having to prove that emissions reductions would not have been achieved without the offset. Others, such as the World Bank, talk of a "flight to quality" as offset providers in voluntary carbon markets increasingly emulate the use of governance and quality assurance tools in compliance markets such as project design documents, third party verification and use of voluntary standards. Reviews of carbon markets observe:

> Over the past 2 years numerous writers and analysts have likened the voluntary carbon markets to the "wild west". In 2007 market trends highlight that this frontier has become a settlement zone. Customers are increasingly savvy about the opportunities and pitfalls in the carbon offset domain and stakeholders are aggressively working to forge the rules of the game and structures to enable smooth transactions (Hamilton et al 2008:53).

As many as 50% of the transactions conducted in 2007 involved credits verified to a specific third party standard. This is clearly a rapidly evolving market.

From being sold as a response to climate change that implied lower transaction costs, it was perhaps inevitable that, as with all markets, property rights and rules are required to bring them into being and ensure their smooth functioning. As Polyani (1944) showed, laissez-faire approaches to markets often produce demands to re-embed markets in frameworks of social control. Even key participants in carbon markets acknowledge the inevitability of this. Abyd Karmali, Managing Director, Global Head of Carbon Markets, Merrill Lynch reflects that:

> Those who assume that the carbon market is purely a private market miss the point that the entire market is a creation of government policy. Moreover, it is important to realize that, to flourish, carbon markets need a strong regulator and approach to governance. This means, for example, that the emission reduction targets must be ratcheted down over time, rules about eligibility of carbon credits must be clear etc. Also, carbon markets need to work in concert with other policies and measures since not even the most ardent market proponents are under any illusion that markets will solve the problem (ClimateChangeCorp 2009).

But beyond the issue of the rules and regulations which should underpin the functioning of carbon markets, other contributors to this book focus on the scalar politics of the new carbon economy and the specific governance challenges it generates (eg see Boyd 2009). Bailey et al explore the politics of scale in relation to the EU Emissions Trading Scheme (ETS) and its connections to the more "global" offset market. It is certainly the case that we find interesting—possibly unprecedented—combinations of public, private and hybrid governance encountering one another and having to work together to enable the new carbon economy to function. Bailey et al show this in relation to the CDM with the armies of auditors, project developers, regulators, lawyers and accountants that have to be mobilised to bring a project into being and see it through the labyrinthine processes of accreditation and verification. It is perhaps the combination of these forms of governance across scales, as well as the governance deficits this often leaves, that create a "wicked" challenge in relation to the new carbon economy.

The response to the question of newness when posed in relation to governance might focus on the plurality of governance forms or modes, the multiple levels at which it operates and the process challenges that are intensified when an attempt is made to govern globally (as with the CDM Executive Board) local resource use decisions around forests, waste and energy, imbued as they are in conflicting systems of value and property rights in diverse settings. This comes through clearly in David Lansing's contribution which shows what happens when the logics of global capital and local socio-ecological systems meet through the new carbon economy. The potential for negative social and environmental consequences is a function of the "articulation between the abstract representations required of commodification and the socio-ecological complexity of locally produced natures". Again, fair trade or other supply chains in commodities like seeds, timber or coffee require elaborate systems of tracing and tracking, attempts to monitor local conditions by global actors (buyers and multinational companies) and enroll "local" resources in "global exchange". Perhaps what is unique in relation to the compliance carbon markets is the UN's role as an arbiter of quality control and the public good property which attempts to commodify carbon must also demonstrate.

Another interesting theme in the book is not just how the new carbon economy creates particular governance challenges and requires specific forms and practices of governance in order to work, but how distinct governance systems intended for other purposes are incorporated into the new carbon economy and have to be re-worked to ensure they can operate in the service of carbon trading. These include the activities of credit-rating agencies, the creation of insurance products (on volumes of CERs likely to be delivered for example), (weather) derivatives, systems of disclosure (such as the Carbon Disclosure Project): what Descheneau and Paterson (this volume) refer to as the "routinization" of financial products in carbon markets. In this way, "carbon market actors borrow from existing financial practices to make the emerging market readily intelligible, to enable it to operate as a matter of financial routine". Interestingly their chapter also highlights the importance of "desire" whereby what is being sold is "not the tonne per se but rather the financial or discursive representations of it".

Claims of newness might also focus on the technologies of governance or governmentalities that have to be employed to make the new carbon economy work through elaborate systems of auditing, measurement and accounting, as Lovell and MacKenzie show so vividly (this volume). Put more critically, Prudham, echoing Lansing's claims, suggests these practices "render the messy materiality of life legible as discrete entities, individuated and abstracted from the complex social and ecological integuments" (Prudham 2007:414). Again, the problem of commensurability is not unique to carbon markets. Money-based economies require us to place "comparable" values on products as diverse as water, coffee and bananas, for example. The price paid does not accurately reflect the labour value or environmental costs invested in their production and it requires an act of faith on the part of consumers to value them through the medium of money and price determined by supply and demand. This draws our attention again to the point Descheneau and Paterson make about the importance of the rituals and imagery that is used to create "romance" around the new carbon

economy and the products and services it seeks to sell in the production
of desire. Brand and image management are crucial to packaging a
particular carbon product to a client wanting to gain public relations
value from it.

Effects of the "New" Carbon Economy

It is perhaps as a result of both the material(ity) of carbon, its value
and omnipresence, that underpins claims about the "nature" of the
new carbon economy, and the forms of governance produced by and
demanded of the new carbon economy, that create in their wake a series
of social, environmental and political effects. This gives rise to a third set
of claims about novelty which the contributions to this book touch upon.

There are claims of the uneven development that results from the ways
in which carbon is commodified or offsets produced in the chapters by
Bumpus and Gutierrez. Again, claims about the uneven (and combined)
nature of development in a capitalist context have a very long history
(Callinicos and Rosenberg 2008). In the context of the new carbon
economy such claims refer to the inequalities in consumption and
resource access which underpin carbon markets by making it more
cost-effective and lucrative to pay for emissions reductions in poorer
parts of the world. Gutierrez highlights the role of time and risk in
particular in reproducing uneven development. Such claims also refer
to the additional value which resources such as forests or municipal
waste can attract as a result of carbon markets—forests because they
absorb carbon, waste because methane can be captured and burned and
a vast number of carbon dioxide equivalent credits acquired. Those
living in forests or alongside waste dumps find their livelihoods and life
chances disrupted by the value these resources acquire, which means
forest conservation for carbon sequestration may trump sustainable use
and keeping a waste site open trumps campaigns to close it because of
the finance secured through carbon markets.

Perhaps it could be argued that all interventions in the form of
aid or investment have the potential to reinforce existing inequalities
and change the value and viability of existing livelihood options, not
just carbon markets. Carbon markets may be unique in their scope
in that gases that are not carbon but which contribute to climate
change, and which emanate from a vast array of human activities, are
'made' equivalent. This opens up nearly all sectors of the economy
in all parts of the world to the potential reach of carbon markets. The
anonymity of the exchange and the distance between buyer and seller
in carbon markets may be greater than in direct negotiations with
donors or investors, a fact again which heightens both the importance of
effective mediating institutions and the work of imagery and narrative

in constructing convincing stories which connect abstract commodities to particular places.

Beyond the material and economic consequences of the carbon economy, we also observe in the chapter by Bailey et al the political consequences of the preference to address climate change by producing markets in carbon. These authors explore the ways in which notions of ecological modernisation are invoked to rationalise and justify the enhanced marketisation of environmental governance to the exclusion of adequate consideration of other policy alternatives. Equally viable options in technological or even economic terms get screened in or out of policy debate because of their political acceptability. Phadke's chapter on "big wind" in the USA also shows us, however, that policy responses considered to be important and effective in efforts to address climate change, such as the promotion of renewable energy, also provoke acts of resistance when handled poorly.

Conclusions

The chapters in this book raise a number of challenges for political engagement with the new carbon economy and for academic attempts to make adequate sense of it.

Firstly, there are a set of strategic issues about whether and on what terms to engage with the new carbon economy through efforts to improve its governance, to find spaces to promote alternative policy options or to resist the further development of the new carbon economy. We have seen above how the nature and organisation of the new carbon economy raises a series of issues of equity and (social and environmental) justice about who gains and who bears the costs of responding to climate change in this way. Critics argue that the pursuit of carbon markets as the preferred option has lost us more than 10 years in the battle to keep climate change within tolerable levels. The time required to create them, manage them and deal with the problems they inevitably raise (additionality, fungibility) has meant that 10 years on we have poorly performing carbon markets (from the point of view of seriously reducing emissions) and other alternatives that could have been more effective have been successfully sidelined.

As the chapter in this book by Bailey et al shows, discourses of ecological modernisation and eco-efficiency provide legitimation to claims that markets can be an essential component of responses to climate change while underplaying some of the contradictions and tensions that relying on them implies. And yet we are left with the dilemma of capitalism of one form or another providing the near-term context in which we have to respond to climate change. Mobilising the influence of powerful fractions of capital means identifying viable accumulation strategies that are compatible with the goal of de-carbonisation. Pricing and carbon trading mechanisms are, in the end, just one small component of a much broader transformation that is required in capitalism if the worst effects of climate change are to be averted (Newell and Paterson 2010).

Secondly, at a theoretical level we have seen how concepts and tools drawn from Marxism, post-structuralism, cultural political economy and actor network theory generate insights into specific features and aspects of the new carbon economy. Future work might explore fruitful and productive combinations of these approaches that will produce the sort of multi-faceted and multi-dimensional explanations we need of how the new carbon economy functions and for whom, paying attention all the while to the broader social, economic and ecological relations of which it is part and which, if it is to be successful, it has to transform.

Each of the perspectives utilised in the contributions here produces a different angle on what might or might not be novel about the new carbon economy. Are we witnessing a routine attempt by the social forces of capital to render the challenge of climate change non-threatening to, and even profitable for, its accumulation objectives, or is there evidence of deeper processes of transformation at work? What limits are suggested by the nature of carbon itself or the technologies of governance it requires to be pressed into the service of international responses to climate change or accumulation strategies? How do people resist, engage with and imagine the new carbon economy? These are just some of the questions the chapters here pose, but our aim is ultimately to provoke further debate and reflection about the new carbon economy and what challenges it poses for activists and scholars alike.

Acknowledgements

We are grateful to Professor Diana Liverman for her support in hosting the workshop at Oxford University, which forms the basis of this book. We would also like to thank the participants at that event and at the panel organised on this theme at the 2009 *Association of American Geographers* in Las Vegas, Nevada for their time and insightful contributions. Last, we thank the editors of *Wiley Blackwell* and *Antipode* for their comments, suggestions and guidance.

References

Bachram H (2004) Climate fraud and carbon colonialism: The new trade in greenhouse gases. *Capitalism, Nature, Socialism* 15(4):10–12
Bakker K (2004) *An Uncooperative Commodity: Privatizing Water in England and Wales.* Oxford: Oxford University Press
Boyd E (2009) Governing the Clean Development Mechanism: Global rhetoric versus local realities in carbon sequestration projects. *Environment and Planning A* 41(10):2380–2395
Bridge G (2010) Resource geographies I: Making carbon economies, old and new. *Progress in Human Geography* DOI: 10.1177/0309132510385524

Callinicos A and Rosenberg J (2008) Uneven and combined development: the social-relational substratum of 'the international'? An exchange of letters. *Cambridge Review of International Affairs* 21(1):77–112

Capoor K and Ambrosi P (2008) *State and Trends of the Carbon Market 2008*. Washington DC: World Bank

Capoor K and Ambrosi P (2009) *State and Trends of the Carbon Market 2009*. Washington DC: World Bank

ClimateChangeCorp (2009) Is carbon trading the most cost-effective way to reduce emissions? http://www.climatechangecorp.com/content.asp?ContentID=6064 (last accessed 2 April 2009)

Friends of the Earth (2009) *Sub-Prime Carbon: Re-thinking the World's Largest New Derivatives Market*. Washington DC: Friends of the Earth

Giddens A (2009) *The Politics of Climate Change*. Cambridge: Polity

Gilbertson T and Reyes O (2009) Carbon trading: How it works and why it fails. *Critical Currents* 7 http://www.carbontradewatch.org/articles/new-book-exposes-dangers-of-carbon-market-ahead-of-bolivia-climates.html (last accessed 21 July 2010)

Hamilton K, Sjardin M, Marcello T and Xu G (2008) *Forging a Frontier: State of the Voluntary Carbon Markets 2008*. Washington, DC: Ecosystem Marketplace and New Carbon Finance

Lohmann L (2005) Marketing and making carbon dumps: Commodification, calculation and counter-factuals in climate change mitigation. *Science as Culture* 14(3):203–235

Lohmann L (2006) *Carbon Trading: A Critical Conversation on Climate Change, Privatisation and Power (Development Dialogue 48)*. Uppsala: Dag Hammarskjöld Foundation

Newell P and Paterson M (1998) Climate for business: Global warming, the state and capital. *Review of International Political Economy* 5(4):679–704

Newell P and Paterson M (2009) The politics of the carbon economy. In M Boykoff (ed) *The Politics of Climate Change: A Survey* (pp 80–99). London: Routledge

Newell P and Paterson M (2010) *Climate Capitalism: Global Warming and the Transformation of the Global Economy*. Cambridge: Cambridge University Press

Polyani K (1944) *The Great Transformation: The Political and Economic Origins of our Time*. Boston: Beacon

Prudham S (2007) The fictions of autonomous invention: Accumulation by dispossession, commodification, and life patents in Canada. *Antipode* 39(3):406–429

UNFCCC (2010) CDM http://cdm.unfccc.int/Statistics/index.html (last accessed 13 December 2010)

World Bank (2010) *Development and Climate Change (World Development Report)*. Washington DC: World Bank

Chapter 2
The Matter of Carbon: Understanding the Materiality of tCO$_2$e in Carbon Offsets

Adam G. Bumpus

Introduction

Carbon offsets exist as a new socio-ecological interface in the management of the environment and economy. Offsets are tools to manage anthropogenic climate change and, in some cases, contribute to international sustainable development (Bumpus and Liverman 2008; Liverman 2009; Lovell and Liverman 2010). In addition to supposedly cheap, fast carbon reductions, offsetting also speaks to ethical debates when offsets allow those who can afford it to continue to pollute (Boyd 2009; Lohmann 2006; Lovell, Bulkeley and Liverman 2009; Vandenbergh and Ackerly 2008). Key to certain offsets is their promotion as channels of finance for local sustainable development in developing countries (Brown et al 2004; Bumpus and Liverman 2008; Figueres 2006). Offsets are complex, and centre around the creation of emissions reductions that are then traded as tonnes of carbon dioxide equivalent (tCO$_2$e) on international markets (Bailey and Maresh 2009; Hepburn 2009). In contrast to emissions allowances that are allocated and either given away or auctioned by governments in cap and trade systems, offsets employ specific technologies or forestry mechanisms to reduce emissions in specific project activities. The material differences between different offset project types are significant because they enable and constrain the social relations necessary for offset production (Bakker and Bridge 2006:21). These differences have relevance for use of offsets

The New Carbon Economy, First Edition. Edited by Peter Newell, Max Boykoff and Emily Boyd.
© 2012 Adam G. Bumpus. Book compilation © 2012 Editorial Board of Antipode and Blackwell Publishing Ltd.

as effective instruments to reduce climate-forcing greenhouse gases
and promote sustainable development (Bumpus and Cole 2010; Olsen
and Fenhann 2008a; UNFCCC 2001). Importantly, offsets also have
relevance for long-running debates and scholarship exposing dialectic
interactions in nature–society relations (Bakker 2005; Bridge and Jonas
2002; Castree 2003b, 2008a; FitzSimmons 1989; Robertson 2006).

I approach carbon reductions in offsets as a dynamic, two-way
relationship of mutual influence and adjustment between social systems
and material nature (Bäckstrand and Lövbrand 2006; Bakker and Bridge
2006; Boyd, Prudham and Schurman 2001; Castree 2003b, 2005:155,
160; FitzSimmons 1989; Goodman 2001). The material and discursive
aspects of carbon commodification are interwoven (Bridge and Smith
2003), thus this analysis aims to show how dialectical socionatural
relations influence offsets in the evolution of the new carbon economy
more generally (Baldwin 2009; Boykoff et al 2009; Braun and Castree
1998; Bumpus and Liverman 2008; Escobar 1996; Lohmann 2006;
Oels 2005; Redclift 2009). As others have shown, the significance of
the biophysical world in nature-based industries has to be balanced
with embracing aspects of nature's social production (Boyd, Prudham
and Schurman 2001; Castree 1995; Goodman 2001; Le Billon 2001;
Prudham 2003): capital metabolises nature into exchangeable values
affecting spaces, environments and social relations across place and
scale (Blaikie and Brookfield 1987; Bryant and Bailey 1997; Bryant and
Goodman 2004; Sneddon 2007; Swyngedouw 1999). Others have shown
how the biophysical properties of the non-human world also often resist
these processes of commodification (Bakker 2005); firms must adapt
given the technologies available, their ability to mobilise resources
and local socio-ecological relations that allow nature's articulation
within capitalist processes (Bakker 2003; Castree 2008b:145). In carbon
offsets, broader regulatory systems, governing mechanisms, institutions
and "tactics" are all present to manage the conflicts and contradictions
in the commodification of carbon (Bakker 2009; Callon 2009; Lohmann
2009; Lovell and Liverman 2010:3; MacKenzie 2009).

I focus here on the calculability of tCO_2e through an examination
of the major material dimensions of offset technology deployment
(Le Billon 2001), its specific local social relations, and the role of
tools to govern reductions through carbon standards (Lohmann 2009;
MacKenzie 2009). The analysis concerns the material interaction of
the different *technology* types with local socionatural conditions and
the calculation activities for offsetting that encourage the material and
discursive components of the commodification of carbon. Rather than
aiming for an *integration* of structural Marxist approaches with poststructural accounts (cf Castree 2002), I aim to provide an account of
carbon offsets that finds a more epistemological middle ground that

realises the importance of complementary lenses to the problem through dialectical understanding (Sneddon 2007:170). By better understanding the material dimensions of how carbon offset technologies interact with the environment and local social relations (Lovell and Liverman 2010), we are better able to understand the linkages between the political economies and evolution of the international carbon markets, and local development implications (Bumpus 2009); important critical and practical components for the emerging carbon economy (MacKenzie 2009).

Carbon dioxide is the most common anthropogenic greenhouse gas (IPCC 2007). The calculation of tonnes of carbon dioxide *equivalent* (ie tCO_2e) arises from the need to develop a common benchmark of the global warming potential of the six greenhouse gasses over a 100-year span by using the global warming potential of 1 tonne of carbon as a baseline indicator. This article uses two carbon mitigation projects as case studies of energy-based carbon offsets in Honduras to add empirical depth to the paper, focussing on the creation of tCO_2e for project-based (ie not sectoral or programmatic) offsets: a renewable energy project (small-scale hydroelectric facility), and a biomass efficiency project (improved cookstoves—ICS) project.[1] As part of a broader study on the evolution of carbon offset markets, key informant interviews were conducted with carbon financiers, project developers, verifiers and communities associated with the projects. Document analysis of the case studies and direct observation triangulated the outcomes and conclusions presented here.

This analysis has two principal aims: to develop an analysis of the commodification of carbon in order to explain the socionatural processes of creating a tonne of carbon dioxide equivalent (tCO_2e) and to open up the dialectical tension between the international carbon market and local socionatural relations, mediated by technology type, drawing links between governance of international carbon mechanisms and "local development" through the application of specific offset technology.

Following an explanation of my approach to the analysis, I provide a technical description of "what carbon" is to be commodified in carbon offsets and an outline of the relationship between fundamental concepts of offsetting and project types. I then delve deeper into the processes of commodification used to create carbon reduction credits in order to develop both the theoretical and practical issues for understanding tCO_2e. Drawing on two different carbon offset case studies, I illustrate the major dimensions of materiality that affect their commodification of carbon reductions, and how such materiality links international carbon market evolution to local socionatural conditions. I conclude with the theoretical and policy insights that can be drawn from the analysis.

Making Carbon Reductions

Commodity status is not intrinsic, but "is the result of conscious and unconscious actions of people in specific circumstances" (Castree 2003b:283): global capitalist processes shape localities and the transformation of nature into commodities. At the same time the specific features of the raw materials themselves affect how these processes are undertaken and reworked at multiple scales (cf Barham and Coomes 1994; Swyngedouw 1999). This broad conceptualisation is useful for analysis of carbon reductions in offsets given the broader evolution of the carbon markets and the political economy of offset governance (Bumpus and Liverman 2008), and the local material specificities, and social relations, of technology deployment that allow carbon reductions to be created (Lovell and Liverman 2010).

Carbon offsets rely on "baseline-and-credit" trading systems that "create" assets: tonnes of carbon dioxide equivalent (tCO_2e; Figure 1). These assets (carbon credits) represent the additional carbon reductions from a baseline of emissions through the investment in emission reduction projects that would not have otherwise taken place (Yamin 2005:30). This is the fundamental notion of environmental "additionality" that differentiates the emissions produced by an offset project from the "business-as-usual" scenario of baseline emissions without the project (Michaelowa 2005). These counterfactual scenarios are determined through analyses of the socio-political economic situations in which the offset project is taking place, and the construction of hypothetical future baseline emissions scenarios.

The creation of tCO_2e relies on the implementation of project activities and the processes of calculating, justifying and verifying emissions reductions. The materiality of the carbon, and the real world context in which it is reduced, must be kept in tension with

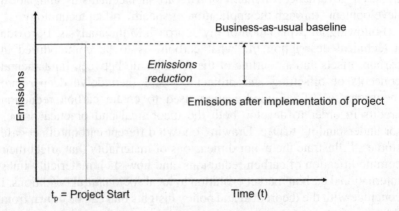

Figure 1: Principles of the baseline determination and the calculation of carbon credits (source: Bumpus and Liverman 2008)

the institutional requirements—such as carbon standards—that exist specifically to assert that a reduction has taken place against the baseline scenario. For energy-based offsets, additionality, baselines and methodologies are developed through the relationship between the demands in the carbon market and the differential calculability of certain technologies to reduce carbon emissions.

The material nature of the technology's engagement with the atmosphere, therefore, plays a crucial role in the effective commodification of the tCO_2e and its ability to be incorporated into carbon standards of differing levels of rigour. Industrial gas destruction of hydrofluorocarbon-23 (HFC-23), for example, is materially much easier to measure and define compared with, for example, afforestation and reforestation (A&R) projects where baselines of tree growth, permanence of carbon sequestered and monitoring requirements are more difficult to assert. As a result industrial gas destruction has been easily incorporated into the legal requirements of the Kyoto Protocol's Clean Development Mechanism (CDM) and is, therefore, subject to rigorous accountability. On the other hand A&R has historically been incorporated into voluntary carbon offset (VCO) markets[2] where, because of the voluntary (ie not legally binding) nature of the market, less rigorous carbon reductions are balanced by, for example, public demands for forest projects (Bayon, Hawn and Hamilton 2007; Hamilton et al 2008; Liverman and Boyd 2008; UNEP-Risoe 2010).

Carbon offsets have some very specific attributes associated with their commodification that contrast them to commodification in other "natures". The most important of which is that, in contrast to commodifying a unit of nature in order to govern its existence, like timber, carbon offsets create a commodity and value out of a piece of nature—carbon dioxide in the atmosphere—that, if achieved properly, *does not exist*. Carbon offsets in their entirety can, therefore, be considered to consist of four interrelated "forms" of carbon (see Table 1): the carbon that continues to be emitted by the offset credit buyer (type 1); the carbon that would have been emitted if it had not been displaced by the project activity (type 2); the lower emissions as a result of the project activity (type 3); and the tCO_2e (type 4) that is produced by the difference in emissions as a result of the project activity and the baseline. A simple equation describes the relationship between these types:

$$\text{type } 2 - \text{type } 3 = \text{type } 4$$

And under ideal hypothetical conditions:

$$\text{type } 1 = \text{type } 4$$

The important point to note for this analysis is that the material conditions of all "types" of carbon associated with an offset project must be considered. Given these conditions, which exist in multiple

Table 1: Description of the four types of carbon embodied in the case studies presented here[11]

Offset project technology	Type 1 carbon: carbon still emitted by buyer of tCO_2e	Type 2 carbon: baseline emissions	Type 3 carbon: displaced carbon	Type 4 carbon: tCO_2e produced by project activity
Grid-connected hydroelectricity project	Industrial emissions in Europe above a regulated target (ie incentive to buy offset credits)	Diesel; emissions from fossil fuel combustion that would have been emitted in the absence of the project	Emissions from fossil fuel burning after project implementation (ie lower emissions when hydroelectricity fed into grid)	Difference in emissions as a result of the project activity compared with the baseline. Hydroelectric output measured at source.
Decentralised Improved Cookstoves	Emissions from clients buying offsets voluntarily for public relations/ marketing[12]	Non-renewable biomass burnt in traditional three-stone fire cookstoves	Reduction in use of non-renewable biomass through the use of more efficient cookstoves	Difference in emissions. Statistical sampling of fuelwood burnt by families with and without improved cookstoves

tCO_2e, tonnes of carbon dioxide equivalent.

locations, over varying timeframes and with different actors, carbon reductions must be conceptualised relationally: it is only within the historical, material and social contexts in which it exists that we can understand "what" is the carbon we are reducing, how is it being reduced (if at all), who stands to benefit from its commodification and with what consequences.[3]

Creating the Carbon Commodity: Processes of Commodification and the International Carbon Economy

The process of creating a carbon commodity exists within a constant dialectical tension between the international carbon market and local socionatural relations. tCO_2e is commodified through socio-technical processes that govern the categorisation of carbon reductions; a process that is strongly mediated by the type of technology used to displace emissions and create reductions.

The creation of tCO_2e is governed by the underpinning principles of offsets that aim to guarantee emissions reductions, and therefore their validity in a market for climate change mitigation. Generally accepted principles of carbon offsets are that they are real, additional, permanent and verifiable (Broekhoff and Zyla 2008). In order to materially attest these principles, carbon standards require documents and processes that define the carbon reductions in offsets. These processes are often dictated at a distance, created by actors outside of the local site of carbon reductions and are created to assist in the transformation of reductions into tradable credits. This section briefly sketches the commodification processes in carbon offsets in order to understand how these processes bind offsets to multiple actors and locations through technology. The section focuses on "constructing the carbon commodity", drawing, where appropriate, on analyses of specifying commodity processes in nature (Boyd, Prudham and Schurman 2001; Castree 2003; Robertson 2000, 2004, 2006). The analysis necessarily engages the role of carbon standards in defining a tCO_2e; it does not provide a specific comparison between standards and instead assumes that the principles of offset standards are a useful heuristic for understanding tCO_2e commodification. I show here that tensions exist between the ability to determine material reductions and the requirements of the market that govern processes of disciplining and holding emissions reductions in place to enable their commodification.

Privatisation in Carbon Offsets

Under market environmental modes of governance (Liverman 2004), emissions reductions have to be assigned rights of ownership—the assignation of legal title to a named individual, group or institution—so that they can be traded as commodities allowing future exchange (Castree 2003b). For offsets to generate future carbon reductions, forward contracts are negotiated through Emissions Reduction Purchase Agreements (ERPAs; Yamin 2005), discursively and legally privatising the tCO_2e that the project is predicted to generate.[4] Following the creation of the project idea notes and preliminary feasibility studies, the ERPA begins a process of legally linking two (or more) actors together in carbon reductions and credit purchasing. As a minimum, an ERPA defines the property rights of a commodity's first exchange and trade from the organisation(s) generating the credit to the investor who lays claim to it.[5] The privatisation of the communal atmosphere through creating purchase agreements and quantifying carbon reductions arising from project activities thus provides control of the commodity to buyers (a potential "allowance to emit") and embodies the legal and discursive privatisation of carbon in the commodification process (Castree 2003b).

Individuation and Abstraction in Carbon Offsets

The carbon commodity created in offsets relies solely on its codification and categorisation by experts as a result of analyses of project activities and baseline scenarios. This process of categorising and separating out an entity or specific thing from its supporting context is known as "individuation" (Castree 2003b:280). In project-based carbon offsets, carbon reductions are individuated and functionally abstracted through a representational and physical (discursive and practical) cut to create units of nature that are deemed socially useful (ie credits that represent a tonne of emissions reduction; Bakker 2005; Mansfield 2004; Robertson 2006). I use the notion of "hemming in", defined as "to confine or be bound by an environment of any kind: to enclose, shut in, limit, restrain" (Oxford English Dictionary 2009), in order to represent the notion that the porosity or movement of the credit is restricted by the process that aims to define its existence (see Figure 2).

Individuation and projected functional abstraction in offsets is created through the a priori representational separation of a current emissions trajectory from a hypothetical one through calculations and assumptions that are necessarily complex (Lohmann 2009). In this way, energy-based offsets do not constitute a *reversal* in emissions; they simply slow the path of *potential* emissions. The Project Design Document (PDD), and its constituent components—methodologies for calculating baselines, proving additionality and the monitoring plan to ensure effective monitoring of future emissions reductions—are all present to constrain, define and individuate carbon reductions *before* they

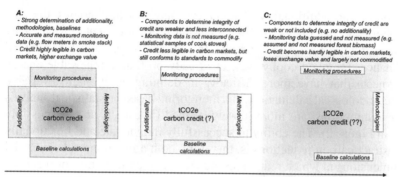

Figure 2: A conceptual diagram illustrating the "hemming in" of carbon credits through the practices of determining project additionality, baseline calculations, methodologies and monitoring procedures. Carbon credits as tCO_2e are individuated from the atmosphere when these four practices determine the material reduction of a tonne of carbon from the atmosphere. Projects that have a material basis that allows these components to be easily defined create a more certain commodity (A). As projects become more complex and technologies less well understood, accurate accounts of these four components can become weaker (moving from A to C). The commodification of the tonne of carbon dioxide equivalent becomes more difficult and finally the integrity of the carbon credit as a saleable commodity on the market becomes difficult to assure (C) (source: Bumpus 2009)

have been reduced. Baselines are, therefore, spatiotemporally context-dependent, specific and variable. This situation has serious ramifications for climate policy, where reduction commodity potential is easily *ex ante* created or destroyed based upon calculation of future baseline scenarios. In this way, the abstraction of carbon "ties the commodification of nature to systems of representation more broadly . . . as regimes of calculation and expertise that more generally make nature and territory 'legible' and governable" (Prudham 2009:130). These processes are the first step in materially creating a tCO_2e, which, given its non-existential nature, could be termed a piece of "counterfactual material nature".[6] Carbon reduction processes are therefore similar to specific other natures, such as the commodification of wetlands, and rely on an "act of reference" (Robertson 2000:472) in order to convey them, albeit imperfectly, in a market. The ability of carbon commodification processes to effectively "hem in" the carbon reductions changes with the different specific material dimensions of reducing carbon: some projects contain more "uncooperative carbon" (following Bakker 2005) than others, where "hemming in" is more difficult and the commodification process harder to complete. The importance of this concept is that specific material dimensions of certain projects and technologies make some tCO_2e more tangible than others. For example, projects such as HFC-23 reduction, although criticised (Wara 2007), actually provide more clear additionality than others, have methodologies that are robust and calculable, and have carbon reductions that can be easily measured (ie case A in Figure 2). In contrast, projects such as blended finance improved cookstoves, or already profitable renewable energy projects that have more difficulty in defining additionality, variable methodological calculations and monitoring based on statistical analysis or assumptions, are more loosely defined (ie case B or even possibly C in Figure 2). The ability to successfully individuate and functionally abstract carbon reductions is, therefore, both a function of the material dimensions in which reductions are created and the ability for practices and documents to effectively assert those reductions.

All offset projects can, therefore, be seen as existing on a spectrum of "more" or "less" uncooperative carbon. An example of this is the bottleneck and difficulties in passing methodologies under the CDM and the problem of defining emissions reductions from different projects to a standardised level (Sterk and Wittneben 2006; World Bank 2006a). As Morgan Robertson points out, there is a "problem of measurement" (2004:367), which, in this case, is highly determined by the technology implemented and the socionatural context in which it is used. This affects the extent to which carbon can be "hemmed in" in an increasingly cognisant carbon market that requires certainty that carbon reductions are real (Gunther 2008).

This tighter "hemming in" has meant that capital accumulation in carbon offsets has been closely tied to firms' differential ability to work with a specific technology, create new forms of governance for reductions, and more accurately determine carbon reductions. It has also been responsible for the call for serious reform of the CDM because of the inherent difficulties with understanding the material application of methodologies and additionality, promoting more sectoral reductions which rely less on such "hemming in" on a project-by-project basis (Figueres 2004; Olsen and Fenhann 2008a; Sterk and Wittneben 2006; Wara and Victor 2008).

Monitoring and Verification: Actual and Spatial Abstraction
Carbon reductions are spatially abstracted as information in order to continue their commodification and placement into wider systems of exchange. Robertson (2000) notes that spatial abstraction essentially involves treating an individualised thing in one place as essentially the same as any apparently similar thing located elsewhere. This process across space belies subtle, and unavoidable, differences in commodities and the contexts in which they were produced. Spatial abstraction in offsets allows credits to flow from the place of carbon reduction to the place of carbon emissions (the fundamental rationale of using offsets)[7]

and allows them to be commensurable with both emissions (Type 1) and other credits in the same system produced under different local socionatural conditions.

The material abstraction in the carbon reduction takes place once the project has been implemented through the execution of the monitoring plan associated with the technology use, and the *ex-post* verification of the reductions by third party organisations. Verification therefore attempts to hem in the credit based on calculating *ex-post* reductions to assert that a material reduction has taken place. Information about the amount of carbon reductions that have been achieved is spatially abstracted when it is conveyed, as data, from the local project site(s), to a third party verifier and then on to registries or emissions reduction accounts of the carbon offset investor, project developer or entity that owns the rights to the tCO_2e.

Through individuation and abstraction, information about a climatic service (cf Thornes and Randalls 2007) is used to justify the creation of a product that can then be sold on and/or used to justify domestic emissions. Without the ability to be spatially abstracted, carbon reduced through offset projects in the global South can never be fungible with the carbon emissions they are supposed to balance in the North: offsets rely on spatial abstraction because the carbon dioxide reduced in one place must be "seen" to be the same as carbon dioxide that is emitted in another. This process allows carbon reductions to be "abstracted from their place specificity" (Robertson 2000:478) and to be displaced into wider systems of exchange, ultimately leading to the credit's retirement, creating the moment when a tonne emitted is actually offset.

Displacement and Exchange in Carbon Offsets

The principal aim of carbon offsets is to provide cheaper emissions reductions across space (Böhringer 2003). Carbon reductions created through the CDM, for example, are used for compliance under the Kyoto Protocol or the European Union Emissions Trading Scheme (EU ETS) and therefore represent a generic licence to emit.[8] This point represents a key tenet of the commodification: the displacement of nature, and its severance from sites of production and specificity.

Displacement can be seen to occur in carbon offsets as credits are abstracted across space as pieces of information, and issued to project proponents from regulatory schemes and voluntary offset buyers. Displacement may obscure the social conditions of its generation where credits are used purely for legal compliance where the local material conditions of reduction are of low importance. This "generic carbon" forms one of Castree's (2003b) sextet of commodification and can easily be seen in the "secondary CDM" market where spot trading of CERs that have already been commodified and issued through the CDM process, are subsequently traded on carbon exchanges. The movement of credits across virtual accounts, through the International Transaction Log of the CDM, "tracks" the credits and aims to avoid double counting (the same tCO_2e being counted more than once by different actors). On the other hand, local conditions may be emphasised in order to gain higher prices in specialist, "boutique carbon" markets (Capoor and Ambrosi 2009), driven in the North in times of excess credits or for buyers looking for development stories for public relations activities in addition to carbon reductions (Hamilton et al 2009; Taiyab 2006). In both cases the carbon market can create demand for specific forms of the carbon commodity, "generic carbon" or "high sustainable development carbon" (Hamilton et al 2009), and uses documentation to describe basic reductions or additional requirements, such as a "sustainable development matrix" in the CDM Gold Standard to cater for market needs. Displacement occurs in all offsets, but is differential, contingent on the carbon markets demanding it.

As a commodity, carbon can then be exchanged and dealt with similar to any other commodity listed on exchanges (as the secondary CDM market shows) and can be retired to compensate for emissions under legal systems, such as the EU ETS (Jepma 2003). Their commodity status affords them a valuation at higher prices than forward contract credits; driven largely by supply/demand and emerging regulatory factors in the carbon market, such as post-Kyoto negotiations and EU legislation on carbon trading (Bailey 2007; Yamin 2005). Offset credits are then finally retired and taken out of exchange circulation in order that the tonne emitted is balanced by the retirement of the tonne reduced (in credit form). In the CDM, the Kyoto Protocol mandates that countries retire credits through official registries to record each tCO_2e. In the VCO market retirement is much more disparately governed, although new registries are emerging to centralise this process and to avoid double counting of emissions (Gillenwater, Sussman and Cohen 2007; Kollmuss, Zink and Polycarp 2008). It is at the point of retirement that the materiality of the commodification of the credit is realised and the final substitution of the tonne emitted with the tonne reduced is rendered materially commensurable (eg type 1 = type 4). Retirement of the credit also acts to "re-veil" the carbon commodity from a "far flung" place (Hartwick 1998; Smith 2007), reaffirming its displacement through consumption (Lovell, Bulkeley and Liverman 2009).

Global–Local Linkages and Offset Technologies
We can see that carbon offsets undergo a complex process of defining emissions reductions in order to create tCO_2e as a tradable commodity.

Offsets create a "counterfactual material nature" in order to place carbon reductions into wider systems of commodity exchange. As I show here, the complicated, contested nature of creating baselines and justifying additionality makes carbon more or less uncooperative: the actual emissions reductions gained from different projects vary according to the process of disciplining carbon reductions. These complex relational processes create a constant tension between the international carbon market and local socionatural relations that is strongly mediated by the type of technology used to displace emissions and create reductions. Table 2 outlines the principal components of each of the case studies, illustrating that although both attempt to produce tCO_2e, they do so via very different material engagements with the atmosphere and processes of defining emissions.

Major material dimensions of the projects affect the ease with which they are incorporated into carbon markets (cf Le Billon 2001). Consequently, there are actions that carbon project developers take to adapt to and manage these uncooperative components of projects in order to fit material conditions into the governance structures, institutional requirements and technical capabilities (Bakker 2009) that carbon markets require (Table 3).

Commodifying Credits from Hydroelectricity

The carbon credits generated by the hydroelectric project were registered under the CDM and used in Europe for compliance under the EU ETS. Legally, a tCO_2e from the project is considered to be fully commensurable with a tonne emitted by a facility covered by the ETS. However, although these form commodities are used for legal compliance, their creation was not without contestation and obstacles. The material dimensions of additionality exemplify the "political life" that abiotic natures (such as carbon) can imbue (Bakker and Bridge 2006). Project developers in Honduras noted that low amounts of up-front carbon finance, and variable carbon prices, did not provide sufficient incentives to invest in capital-intensive projects, such as a hydroelectricity dam. Indeed, the project developers had to argue for *and* against the project's additionality to different investors in order to gain finance for the project. Defining the project's additionality, therefore, relied on the process of justification and negotiation between interested parties. Similar to the creation of other new economic systems through practices of calculation (cf Mitchell 2008), different carbon offset actors introduce different calculations to persuade others that they are superior to rival models and additionality scenarios. In addition to the production of technical documents, the a priori creation of carbon reductions, therefore, also exists through a discursive exercise of power

Table 2: A table outlining the principal components of two case study projects in Honduras and their relative material dimensions (source: Bumpus 2009)

Component	Case study A: Hydroelectricity project	Case study B: Improved cookstove (ICS) project
Project type/overview		
	Run of river, daily containment hydroelectric dam. 13.5 MW Large centralised discrete (1 site) capital project (~$15million)	Improved cookstoves, up-scaling of existing cookstove project (~$50,000) with 1600 stoves (multiple sites)
Creating carbon reductions		
Actual reduction	Displaces diesel electricity generation in national grid (national scale reductions)	Replaces traditional wood-burning stoves with more efficient stoves that burn less non-renewable biomass (household/ community scale reductions)
Documentation supporting commodification	Documents show: legal title over the emissions reductions, baseline of emissions, methodologies to calculate carbon reductions (relies on grid that would have been burning diesel in place of hydro dam, barriers to investment for hydro facilities), verification and monitoring of emissions reductions and movement of "credits" across space. Documents officially sanctioned by UN system and credits transferred through official registries	Documents follow UN process (in small-scale CDM), but not officially sanctioned. No existing documents for methodologies because of complexity of emissions reductions and lack of UN-sanctioned official support for improved cookstove projects at start of project (2005). New documents created by verifiers to produce methodology for calculation and monitoring of emissions reductions. Carbon finance up-scaled operations to allow more stoves to be produced
Monitoring carbon reductions	Centralised project with well understood technology and material carbon reductions led to clear methodologies and guidance on associated emissions reductions for international market	No clear guidance or mandatory governance structures because of difficulty in methodologies and calculations for improved cookstove projects. This has changed over time as lessons learned about the material difficulties of projects are fed into governing standards

Table 2: *(Continued)*

Component	Case study A: Hydroelectricity project	Case study B: Improved cookstove (ICS) project
Link to international markets	Credits flow into regulated markets to comply with EU and Kyoto emissions reduction requirements	Credits used in offset company offering voluntary emissions reductions for consumers/companies outside of the Kyoto process
Biophysical properties for carbon reductions		
Opportunities as result of biophysical constitution	Situated in mountainous region: elevation and water availability. River highly polluted, little opposition or opportunity cost for water, communities amenable to reforesting hillsides. Honduras generating most electricity from diesel therefore opportunity to reduce emissions	Situated in peri-urban areas of capital city in shanty towns: demand for wood high, prices high, therefore more uptake likely (because of increased economic efficiency and lack of available wood to cut and collect by household)
Physical in situ obstacles as a result of biophysical constitution	Needs rainfall. Needs national power company to maintain grid infrastructure otherwise hydro plant cannot feed power into the grid Need to have cooperation from local communities to ensure reforestation of watershed (therefore efficient running of dam)	Stoves degenerate over time (difficult to calculate efficiency); variation in stove use between households and over time. Possible use of different, untested, biomass in combustion. Monitoring of carbon difficult because decentralised, dispersed sites of reduction (household kitchens).
Institutional obstacles as a result of biophysical constitution	Officially none, although questions of additionality remain and future demand for energy which is not outstripped by "clean" hydro	At time of implementation, no approved methodologies under UN system; project developer had to create own systems. Lack of registration to standard affects exchange value of credit

and argument. In this way, the material nature of the dam's expense provided both an opportunity to argue for carbon finance because it was considered rhetorically additional, and an obstacle to creating *actual* material carbon reductions because, in reality, the project may very likely have gone ahead anyway. Despite these difficulties, CDM regulatory bodies easily accepted the additionality of the project, even though the material nature of implementing a multi-million dollar hydro

Table 3: Major material dimensions and responses for the case study carbon offsets (sources: Atmosfair 2008; Gold Standard 2008, 2009; UNFCCC 2005)

Material dimensions	Hydro	Stoves
Measurability of emissions reduction activity (type 3 carbon)	High; measured at source	Low; estimated from statistical samples[13]
Distribution of reductions	Centralised	Decentralised
Measurability of displaced carbon (type 2 carbon)	High; switching off bunker fuels in real time as hydro comes online	Low; fraction of non-renewable biomass difficult to assert, statistical sampling of stove use within different user communities
Inclusion possibilities in carbon markets	Able to be included in compliance markets; material calculations assisted in project being one of the first to be registered	Mostly included in voluntary market, with more recently recognised methodologies incorporated into more rigorous carbon standards
Local responses to deal with material basis of uncooperative carbon	Relatively simple: some technological implementation; restoring watershed	More complicated: improving local governance for monitoring; attends to local understanding of carbon component of project in order to improve information provision

dam project meant that it had to be financially stable with and without carbon finance (for other examples, see Lohmann 2009).

In contrast to potential difficulties in defining additionality, calculating future carbon reductions was relatively simple to justify for the hydro facility because of the nature of the project: electricity generation from water flow replaces fossil fuel used for generation in the grid. The centralised nature of the project meant that predicted energy outputs, and therefore displacement of fossil fuels, could be predicted and then monitored and measured easily, thereby allowing easier individuation and spatial abstraction of the credits produced. This situation existed even though biophysical phenomena, such as excess rain, which caused a generator to explode, or insufficient rain, which meant the generators couldn't run, affected the overall energy output and carbon credit production. Dealing with these obstacles posed by the project's material dimensions, however, was relatively easy: clutches were installed on generators to avoid over spinning, and a comprehensive local reforestation plan was enacted in order to improve the watershed for the facility. Moreover, these material obstacles were not barriers to the project's ability to generate carbon credits; indeed the reforestation activities were actually used as an opportunity to describe positive local effects, assisting the project in selling to high sustainable development carbon markets (Atmosfair 2008; World Bank 2006b, 2008). The "socionatural–technical complexes" that create carbon credits in this situation are, therefore, at once both material and discursive in nature (Escobar 1999; Mitchell 2008), and dynamically connect local conditions in the South to broader narratives and demands of the carbon market in the North (Bumpus and Cole 2010).

Reworking Commodification in Cookstoves

Due in large part to material constraints of including cookstove projects as carbon offsets,[9] carbon finance systems have only recently been recognised in formal systems such as the CDM and Gold Standard (GTZ 2010; Mann 2007). Indeed, obstacles to inclusion in carbon finance existed by virtue of individual stoves' physical interrelation with the environment, and the calculability requirements for bringing stoves into carbon offset mechanisms. For example, additionality was difficult to assert in some of the stoves associated with case study B because of multiple funding avenues to local project implementers from both carbon and non-carbon sources. In addition, the decentralised and widely distributed sites of carbon reductions created obstacles to effective monitoring and understanding accurately over time the amount of carbon reductions achieved. As a manager of the project noted, "we watered down the monitoring requirements because it was a small project and we were running out of time due to [verifier's] other commitments". These difficulties and lower carbon calculations are not surprising given the project's pilot status and early inclusion in the carbon markets when voluntary projects shifted from relying more on consumer-offset connections (Lovell, Bulkeley and Liverman 2009) to the requirements of calculating standards (Hamilton et al 2008).

Although these limited methodological processes can be partly explained by the projects' pilot status, the *inherent* difficulties can be explained through the relationship between the technology's specific engagement with the atmosphere and the inadequate (but evolving) ability of socio-technical systems to define the carbon reductions that delineate the extent to which carbon credits can be effectively commodified. In this case, the carbon reductions were difficult to "disentangle" from their broader socionatural context (Lohmann 2009:509). Two material factors complicated individuation in cookstoves. Firstly, the methodologies for the project cannot *measure* the use of technology that leads to emissions reductions (ie type 3 carbon), but must be made on estimates of samples of stove use that attempt to allow extrapolation to the heterogeneous communities they represent. Secondly, difficulty in calculating the fraction of non-renewable biomass (fNRB; ie type 2 carbon) used in cooking—a crucial factor in determining emissions reductions—also adds another layer of complexity to the accuracy of methodologies (see Table 1) (Bailis et al 2007; Edwards et al 2004; Masera et al 2006; Smith et al 2000:758).[10] This early work to calculate stove carbon reductions, however, did lay the foundations for the development of methodologies and their registration under the Gold Standard (Gold Standard 2009).

In the absence of wide-scale digital monitoring of emissions from individual stoves, improved local inclusion has become important for cookstove offset projects (Climate Care 2009; Harvey 2009; Gold Standard 2009). MacKenzie (2009:451) notes that the evolution of the carbon markets over time may have exposed inconsistencies and flaws, allowing the creation of tCO_2e to consolidate in news forms. Similarly, the evolution of accounting for "uncooperative carbon" has meant that the practices of calculation have also had to evolve. For stoves, this has included increased participation of local actors in both the disciplining of carbon reductions and the evolution of accounting techniques. The difficulty of monitoring has also led to a possible renegotiation of the privatisation of the carbon credits themselves, and the *need* for improved engagement with local institutions to understand the constraints, and possibilities, that carbon commodification places on the project (Bumpus 2009).

Based on the analysis here, and at the risk of oversimplification, Figure 3 provides a basic heuristic interpretation of the dialectical tensions between the carbon market requirements, local material dimensions of reductions and the commodification process as it relates to project type and local socionatural processes. The point to note is that, like other commodified "natures", tCO_2e is created through a constant tension between the processes that aim to define and discipline it and that the possible uncooperative nature that creates obstacles and opens up opportunities to multiple actors (Boyd, Prudham and Schurman 2001). Further research is required to understand how this heuristic is altered for other forms of socionatural technical interactions in other carbon offset types, but serves to illustrate the important tensions that form in the complexes required for the commodification of tCO_2e.

Tentatively, then, complex local socionatural material conditions (where project technologies do not have simple technical solutions to calculate carbon reductions) may allow improved local involvement, and possible benefits from carbon finance. This situation, however, does not inherently attend to the power and bargaining positions between

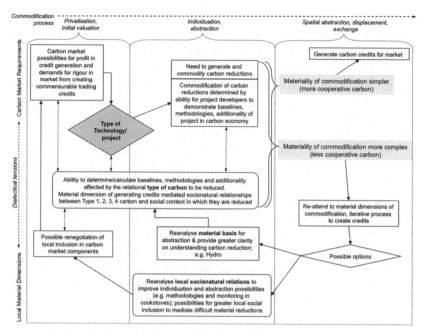

Figure 3: A graphical heuristic to illustrate the dialectical tensions between the materiality of reducing carbon in offset projects and their relations to the global carbon economy through the processes of commodification. In a necessarily simplified process, flow of the process starts in the top left corner with carbon market possibilities. Double-sided arrows represent dialectical tensions between issues (in boxes) and more generally between the evolution of the carbon economy (upper) and the difficulty/importance of the materialities of carbon reductions (lower)

northern carbon capitalists and southern NGOs or groups in need of finance for their development projects (cf Newell, Jenner and Baker 2009), and relative benefits accrued from the project. Continued political ecological research that attends to the political and discursive aspects of carbon credit generation in offsets is required to elaborate on these points.

Conclusion

This piece provides a specific example of socionatural articulations and capital's engagement with the "natural" world. Through offsets, "capitalism produces specific natural environments but these environments are, in turn, both enabling and constraining . . . *in specific relation* to the social relations they are imbricated within" (Castree 1995:24; emphasis in original). The material nature of carbon reductions and the social relations in which these are governed, argued, negotiated and enacted are dialectically related to the broader requirements of the new and evolving carbon economy. Offsets are produced through "socionatural–technical complexes": they are at once both material and discursive, relying on the actions and agency of a number of actors and components of "nature" in order to construct emissions reductions and turn them into tradable commodities. Ultimately the commodification of carbon becomes a political economic, socionatural relational process across a wide range of literal and figurative distances (Lohmann 2009; Pryke, Rose and Whatmore 2003).

The process of commodification of carbon differs in material difficulty according to the engineered technology in question, and the ability for carbon standards to effectively hem in carbon; a process that is inherently difficult. Carbon reductions in offsets, then, are not solely scientifically verifiable material "removals" of carbon from the atmosphere. Instead, behind each carbon credit is a process of commodification that relies on a dialectical relationship between the requirements of the evolving carbon economy, local socionatural– technical complexes, and the relational aspects of connected carbon emissions and reductions (types 1–4).

The limited analysis of case studies here points to the need for more studies that aim to bridge the ontological ground between material and discursive approaches to the creation of the new carbon economy (cf Mitchell 2008). Although carbon offsets can be governed at a distance through "codes of knowledge representing the human interface with the biophysical world" (Baldwin 2009:419), local material dimensions of offsets can redefine the extent to which these techniques are possible. The theoretical analysis here has, therefore, shown that despite the fact that expert advisors in the carbon economy play an "authoritative role in the construction of . . . eco-knowledges", the material difficulties of some technologies, and the specific requirements of markets, may in some circumstances open up possibilities for self-reflection and inclusion of multiple voices in such constructions (Bäckstrand and Lövbrand 2006:54–55). As Lovell and Liverman (2010) note, it is crucial to pay close attention to the material world and the role of technologies, which form a vital part of the "plethora of actors" that make up the broader political economy of carbon offsets. Building on this, I have shown that the commodification *process* intimately connects international governance of carbon finance (and carbon finance mechanisms) with the socio-natural relations in specific places.

Carbon offset technology, therefore, plays a significant role in connecting local processes to global markets and, as such, is a crucial consideration in related debates concerning sustainable development through carbon finance (cf Olsen and Fenhann 2008b) and project-based offset reform more generally. The article has opened up avenues on constructing the carbon commodity as a segue to further understanding geographic interpretations of how carbon offset project locales are wired in (cf Castree 2003a; Massey 1994) to broader political economies and spheres of influence (Simon 2008).

In sum, as tools to mitigate climate change through tangible project activities, offsets sit at junctures between the material properties of reducing carbon emissions, the construction of markets to govern reductions and the social relations that enable offset production. Analysis of the specific connections in these socionatural–technical complexes are important to both the theorisation of socionatures under contemporary and future environmental governance regimes, and policy debates about shaping global environmental change and North/South sustainable development in the twenty-first century.

Acknowledgements

This piece was made possible through an ESRC/NERC studentship at the School of Geography and Environment, Oxford University and a Postdoctoral Fellowship from the Pacific Institute for Climate Solutions (PICS) at ISIS, Sauder School of Business, University of British Columbia. Thanks to the companies involved in the case studies that made field research possible. I would like to extend warm thanks to Professor Margaret FitzSimmons for her extremely insightful discussions and support on this topic during my time as an overseas institutional visitor at the University of California, Santa Cruz in 2008. Professor Diana Liverman also was central in the development of these ideas. Many thanks also to both Professor Peter Newell and Dr Emily Boyd for extremely useful scholarly comments on an earlier draft of this paper, and Dr Alison Shaw and three anonymous reviewers for extremely constructive criticism. All remaining mistakes, of course, remain my responsibility.

Endnotes

[1] The case studies are based on doctoral research undertaken at the University of Oxford (2005–2009) and through extensive fieldwork on carbon offset case studies in Honduras. My aim here is to use the case studies as illustrative of the processes and general tendencies inherent in carbon offsets, rather than applying a critique of their specific carbon accounting per se, or an extrapolation of conclusions to all offset project types and manifestations.

[2] At the time of writing, 14 afforestation and reforestation projects exist in the CDM. However, they have not seen the success of industrial gas reduction, grid connected renewable energy and other centralised technological offset projects in the CDM.

[3] I am indebted to Professor Margaret FitzSimmons for her personal time and valuable insight into these issues in the embryonic stages of this work carried out at the University of California, Santa Cruz in 2008.

[4] Carbon offset projects are also unilaterally developed and then sold into the market. However, forward contracts have been a way of integrating carbon finance into investment calculations to incentivise new project development.

[5] Of course the attraction of carbon trading to traders is that credits can continually be traded until they are finally retired from the system to create the actual offset. However the focus of this research is to understand the processes through which carbon reductions are created, and turned into commensurable commodities in the market— how that carbon is then traded with specific actors after this is the focus of future work in this area (also see Bailey 2007; Lovell and Liverman 2010).

[6] Lohmann (2005) invokes "counterfactual" scenarios in the commodification of carbon in offsets and uses the term to illustrate that emissions reductions are not really created in offsets (see also Byrne et al 2001), whilst others use "counterfactual scenarios" as a technical term to outline various baseline scenarios that would happen in the absence of the offset project (eg Pfaff et al 2000). I invoke the term here to represent the fact that offsets do create questionable carbon reductions in some cases, but that the "material" nature we are dealing with in carbon offsets consists of a "nature" (CO_2) that does not exist (ie counterfactual) yet is still materially bound up in socionatural-technical processes that turn a piece of nature into a commodity through its reduction.

[7] This is especially true of offsets examined here that span North (developed) and South (developing countries). The difference in marginal abatement costs between the North and South is a key driver for the inclusion of offsets (Bumpus and Liverman 2008), although North–North offsets also exist in the form of voluntary offsets in countries without binding emissions targets (eg USA) and in Kyoto's developed country offset mechanism, Joint Implementation.

[8] There are some specific project types, like large-scale hydroelectricity and afforestation and reforestation projects, that are not eligible under the EU ETS, illustrating that carbon is not completely generic within the system, although the variety of projects that do supply the ETS from the CDM are considered as such.

[9] Cookstoves were excluded from the CDM until 2008 for both methodological and categorical reasons that defined improved efficiency of stoves as avoided deforestation, rather than energy efficiency. Despite re-attending to these components of its commodification, difficulties still remain in passing stoves projects in the CDM, with only one having passed registration by May 2010, indicating their difficult incorporation into compliance markets as fully commensurable with other emissions reductions.

[10] Stoves that use renewable biomass are considered to be 'carbon neutral' because the emissions from wood burning are assumed to be sequestered over time through the re-planting of the wood source.

[11] This is a heuristic interpretation to clearly show the relationships between different emissions associated with a project and does not account for baseline changes over time.

[12] Although the buyers of the offsets are doing so for different reasons—one is under legal obligation of the Kyoto Protocol and the others for marketing reasons—offsets are bought under incentives for organisations to reduce their net carbon footprint and thus are still useful in this analysis.

[13] Current methodologies rely on statistical sampling, however there are emerging digital sensor technologies which can be applied to monitor stoves on a continual basis (TWP 2010).

References

Atmosfair (2008) La Esperanza Hydroelectric Project Honduras. Additional annexes to the PDD for retroactive Gold Standard registration. http://www.atmosfair.de /fileadmin/user_upload/Projekte/Honduras/Esperanza_GS_additional_annexes.pdf (last accessed June 15 2008)

Bäckstrand K and Lövbrand E (2006) Planting trees to mitigate climate change: Contested discourses of ecological modernization, green governmentality and civic environmentalism. *Global Environmental Politics* 6(1):50–75

Bailey I (2007) Market environmentalism, new environmental policy instruments, and climate policy in the United Kingdom and Germany. *Annals of the Association of American Geographers* 97(3):530–550

Bailey I and Maresh S (2009) Scales and networks of neoliberal climate governance: The regulatory and territorial logics of European Union emissions trading. *Transactions of the Institute of British Geographers* 34(4):445–461

Bailis R, Berrueta V, Chengappa C, Dutta K, Edwards R, Masera O, Still D and Smith K R (2007) Performance testing for monitoring improved biomass stove interventions: Experiences of the Household Energy and Health Project. *Energy for Sustainable Development* 11(2):57–70

Bakker K (2003) *An Uncooperative Commodity: Privatizing Water in England and Wales.* Oxford: Oxford University Press

Bakker K (2005) Neoliberalizing nature? Market environmentalism in water supply in England and Wales. *Annals of the Association of American Geographers* 95(3):542–565

Bakker K (2009) Neoliberal nature, ecological fixes, and the pitfalls of comparative research. *Environment and Planning A* 41(8):1781–1787

Bakker K and Bridge G (2006) Material worlds? Resource geographies and the "matter of nature". *Progress in Human Geography* 30(1):5–27

Baldwin A (2009) Carbon nullius and racial rule: Race, nature and the cultural politics of forest carbon in Canada. *Antipode* 41(2):231–255

Barham B L and Coomes O T (1994) Reinterpreting the Amazon rubber boom: Investment, the state, and Dutch disease. *Latin American Research Review* 29(2):73–109

Bayon R, Hawn A and Hamilton K (2007) *Voluntary Carbon Markets: An International Business Guide to What They Are and How They Work.* London: Earthscan

Blaikie P M and Brookfield H C (1987) *Land Degradation and society.* London: Methuen

Böhringer C (2003) The Kyoto Protocol: A review and perspectives. *Oxford Review of Economic Policy* 19(3):451–466

Boyd E (2009) Governing the Clean Development Mechanism: Global rhetoric versus local realities in carbon sequestration projects. *Environment and Planning A* 41(10):2380–2395

Boyd W, Prudham S and Schurman R A (2001) Industrial dynamics and the problem of nature. *Society and Natural Resources* 14(7):555–570

Boykoff M T, Bumpus A G, Liverman D M and Randalls S (2009) Theorizing the carbon economy: Introduction to the special issue 41(10):2209–2304

Braun B and Castree N (eds) (1998) *Remaking Reality: Nature at the Millennium.* London: Routledge

Bridge G and Jonas A (2002) Governing nature: The re-regulation of resource access, production, and consumption. *Environment and Planning A* 34:759–766

Bridge G and Smith A (2003) Intimate encounters: culture, economy, commodity. *Environment and Planning D–Society and Space* 21(3):257–268

Broekhoff D and Zyla K (2008) *Outside the Cap: Opportunities and Limitations of Greenhouse Gas Offsets.* http://www.google.ca/url?sa=tandsource=webandct=resandcd=1andved=0CBQQFjAAandurl=http%3A%2F%2Fpdf.wri.org%2Foutside_the_cap.pdfandrct=jandq=WRI+offset+principles+real+permanentandei=UzvrS6zCBIysswO1uOHcDwandusg=AFQjCNH0nfyRGMBdxCZVF8NWk0EZKZD1gA (last accessed 1 June 2008)

Brown K, Adger W N, Boyd E, Corbera-Elizalde E, and Shackley S (2004) How do CDM projects contribute to sustainable development? *Tyndall Centre for Climate Change Research, Technical Report 16.* http://www.tyndall.ac.uk/research/theme2/final_reports/it1_13.pdf (last accessed 30 September 2010)

Bryant R L and Bailey S (1997) *Third World Political Ecology.* London: Routledge

Bryant R L and Goodman M K (2004). Consuming narratives: The political ecology of "alternative" consumption. *Transactions of the Institute of British Geographers* 29(3):344–366

Bumpus A G (2009) "The geographies of carbon offsets: Governance, materialities and development". Unpublished doctoral thesis, University of Oxford

Bumpus A G and Cole J C (2010) How can the current CDM deliver sustainable development? *Wiley Interdisciplinary Reviews: Climate Change* 1(4):541–547

Bumpus A G and Liverman D M (2008) Accumulation by decarbonization and the governance of carbon offsets. *Economic Geography* 84(2):127–155

Byrne J, Glover L, Alleng G, Inniss V, Mun Y M and Wang Y D (2001) The postmodern greenhouse: Creating virtual carbon reductions from business-as-usual energy politics. *Bulletin of Science, Technology and Society* 21(6):443

Callon M (2009) Civilizing markets: Carbon trading between in vitro and in vivo experiments. *Accounting, Organizations and Society* 34(3–4):535–548

Capoor K and Ambrosi P (2009) *State and Trends of the Carbon Market 2009*. Washington DC: The World Bank

Castree N (1995) The nature of produced nature: Materiality and knowledge construction in Marxism. *Antipode* 27(1):12–48

Castree N (2002) False antitheses? Marxism, nature and actor-networks. *Antipode* 34(1):111–146

Castree N (2003a) Bioprospecting: From theory to practice (and back again). *Transactions of the Institute of British Geographers* 28(1):35–55

Castree N (2003b) Commodifying what nature? *Progress in Human Geography* 27(3):273–297

Castree N (2005) *Nature*. London: Routledge

Castree N (2008a) Neoliberalising nature: Processes, effects, and evaluations. *Environment and Planning A* 40(1):153

Castree N (2008b) Neoliberalising nature: The logics of de-regulation and re-regulation. *Environment and Planning A* 40(1):131

Climate Care (2009) Uganda efficient stoves: Project map: Carbon Projects: Reducing Emissions: Low carbon technologies: Climate Care. http://www. jpmorganclimatecare.com/projects/countries/Uganda-efficient-stoves/ (last accessed 22 December 2009)

Edwards R, Smith K R, Zhang J and Y Ma (2004) Implications of changes in household stoves and fuel use in China. *Energy Policy* 32(3):395–411

Escobar A (1996) Constructing nature: Elements for a poststructural political ecology. In R Peet and M Watts (eds) *Liberation Ecologies: Environment, Development, Social Movements* (pp 46–68). London: Routledge

Escobar A (1999) Steps to an antiessentialist political ecology. *Current Anthropology* 40(1):1–30

Figueres C (2004) *Institutional Capacity to Integrate Economic Development and Climate Change Considerations: An Assessment of DNAs in Latin America and the Caribbean*. Washington DC: Inter-American Development Bank

Figueres C (2006) Sectoral CDM: Opening the CDM to the yet unrealized goal of sustainable development. *McGill International Journal of Sustainable Development Law and Policy* 2:5

FitzSimmons M (1989) The matter of nature. *Antipode* 21(2):106–120

Gillenwater M, Sussman F and Cohen J (2007) Policing the voluntary carbon market. *Nature Reports Climate Change*. http://www.nature.com/climate/2007/0711/ full/climate.2007.58.html (last accessed 1 September 2007)

Gold Standard (2008) *An Introduction to the Carbon Credit Protocols from "Methodology for Improved Cook-Stoves and Kitchen Regimes"*. http://www.stovetec

.net/mambo/images/Simple%20Language%20Version%20Gold%20Standard.pdf (last accessed 15 May 2009)

Gold Standard (2009) *Gold Standard Project Registry.* http://goldstandard.apx.com/ (last accessed 25 April 2009)

Goodman D (2001) Ontology matters: The relational materiality of nature and agro-food studies. *Sociologia Ruralis* 41(2):182–200

GTZ (2010) *Carbon Markets for Improved Cooking Stoves—A GTZ Guide for Project Operators.* 3rd ed. Berlin: GTZ-HERA—Poverty-oriented Basic Energy Service

Gunther M (2008) JPMorgan jumps into carbon trading—Aug. 12, 2008. http://money .cnn.com/2008/08/11/technology/jpmorgan_carbon.fortune/index.htm (last accessed 15 May 2009)

Hamilton K, Bayon R, Turner G and Higgins D (2008) *State of the Voluntary Carbon Market 2008.* London: Ecosystem Marketplace New Carbon Finance

Hamilton K, Sjardin M, Shapiro A and Marcello T (2009) *Fortifying the Foundation: State of the Voluntary Carbon Market 2008.* London: Ecosystem Marketplace New Carbon Finance

Hartwick E (1998) Geographies of consumption: A commodity-chain approach. *Environment and Planning D-Society and Space* 16(4):423–437

Harvey A (2009) *Project Design Document Form (GS-VER-PDD): Efficient Cooking Stoves in Uganda.* https://gs1.apx.com/mymodule/ProjectDoc/Project_ViewFile. asp?FileID=1689andIDKEY=f98klasmf8jflkasf8098afnasfkj98f0a9sfsakjflsakjf8dn 2329131 (last accessed 25 September 2009)

Hepburn C (2009) International carbon finance and the Clean Development Mechanism. In D Helm and C Hepburn (eds) *The Economics and Politics of Climate Change* (pp 409–432). Oxford: Oxford University Press

IPCC (2007) Summary for policymakers—Working group 1. In *Climate Change 2007: The Physical Science Basis. Contribution of Working Group 1 to the Fourth Assessment Report of the Intergovernmental Panel on Climate Change.* Cambridge and New York: Cambridge University Press

Jepma C J (2003) The EU emissions trading scheme (ETS): How linked to JI/CDM? *Climate Policy* 3(1):89–94

Kollmuss A, Zink H and Polycarp C (2008). *Making Sense of the Voluntary Carbon Market: A Comparison of Carbon Offset Standards.* Berlin: WWF Germany

Le Billon P (2001) The political ecology of war: Natural resources and armed conflicts. *Political Geography* 20(5):561–584

Liverman D M (2004) Who governs, at what scale, and at what price? *Annals of the Association of American Geographers* 94(4):734–738

Liverman D M (2009) Carbon offsets, the CDM and sustainable development. In H J Schellnhuber (ed) *Global Sustainability: A Nobel Cause* (pp 129–142). Cambridge: Cambridge University Press / Potsdam Institute for Climate Impact Research

Liverman D M and Boyd E (2008) The CDM, ethics and development. In K H Olsen and J Fenhann (eds) *A Reformed CDM* (pp 47–57). Roskilde: Forskningscenter Risø

Lohmann L (2005) Making and marketing carbon dumps. *Science as Culutre* 14(3):203–205

Lohmann L (2006) *Carbon Trading.* Dorset: The Corner House

Lohmann L (2009) Toward a different debate in environmental accounting: The cases of carbon and cost–benefit. *Accounting, Organizations and Society* 34(3–4): 499–534

Lovell H, Bulkeley H and Liverman D M (2009) Carbon offsetting: Sustaining consumption? *Environment and Planning A* 41(10):2357–2379

Lovell H and Liverman D M (2010) Understanding carbon offset technologies. *New Political Economy* 15(2):1–16

MacKenzie D (2009) Making things the same: Gases, emission rights and the politics of carbon markets. *Accounting, Organizations and Society* 34(3–4):440–455

Mann P (2007) Carbon finance for clean cooking—time to grasp the opportunity. *Boiling Point* 54:1–2

Mansfield B (2004) Rules of privatization: Contradictions in neoliberal regulation of North Pacific fisheries. *Annals of the Association of American Geographers* 94(3):565–584

Masera O, Ghilardi A, Drigo R, Angel Trosso M (2006) WISDOM: A GIS based supply demand mapping tool for woodfuel management. *Biomass and Bioenergy* 30(7):618–637

Massey D (1994) *Space, Place and Gender*. Minneapollis: University of Minnesota Press

Michaelowa A (2005) Determination of baselines and additionality for the CDM: A crucial element of credibility of the climate regime. In F Yamin (ed) *Climate Change and Carbon Markets* (pp 289–303). London: Earthscan

Mitchell T (2008). Rethinking economy. *Geoforum* 39(3):1116–1121

Newell P, Jenner N and Baker L (2009) Governing clean development. *Development Policy Review* 27(6):717–741

Oels A (2005) Rendering climate change governable. *Journal of Environmental Policy and Planning* 7(3):185–207

Olsen K H and Fenhann J V (2008a) Sustainable development benefits of clean development mechanism projects: A new methodology for sustainability assessment based on text analysis of the project design documents submitted for validation. *Energy Policy* 36(8):2773–2784

Olsen K H and Fenhann J V (2008b) *A Reformed CDM*. Roskilde: Forskningscenter Risø

Oxford English Dictionary (2009) *Oxford English Dictionary*. http://dictionary. oed.com/cgi/entry/50104624?query_type=wordandqueryword=hem+inandfirst= 1andmax_to_show=10andsingle=1andsort_type=alpha (last accessed 15 May 2009)

Pfaff A S P, Kerr S, Hughes R F, Liu S G, Sanchez-Azofeifa G A, Schimel D, Tosi J and Watson V (2000). The Kyoto protocol and payments for tropical forest: An interdisciplinary method for estimating carbon-offset supply and increasing the feasibility of a carbon market under the CDM. *Ecological Economics* 35(2): 203–221

Prudham S (2003) Taming trees: Capital, science, and nature in Pacific slope tree improvement. *Annals of the Association of American Geographers* 93(3): 636–656

Prudham S (2009) Commodification. In N Castree, D Demeritt, D L Liverman and B Rhoads (eds) *Companion to Environmental Geography* (pp 123–142). Oxford: Wiley-Blackwell

Pryke M, Rose G and Whatmore S (2003) *Using Social Theory*. London: Sage

Redclift M (2009) The environment and carbon dependence: Landscapes of sustainability and materiality. *Current Sociology* 57(3):369–387

Robertson M M (2000) No net loss: Wetland restoration and the incomplete capitalization of nature. *Antipode* 32(4):463–493

Robertson M M (2004) The neoliberalization of ecosystem services: Wetland mitigation banking and problems in environmental governance. *Geoforum* 35(3):361–373

Robertson M M (2006) The nature that capital can see: Science, state, and market in the commodification of ecosystem services. *Environment and Planning D: Society and Space* 24(3):367–387

Simon D (2008) Political ecology and development: Intersections, explorations and challenges arising from the work of Piers Blaikie. *Geoforum* 39(2):698–707

Smith K R, Uma R, Kishore V V N, Zhang J F, Joshi V and Khalil M A K (2000) Greenhouse implications of household stoves: An analysis for India. *Annual Review of Energy and the Environment* 25:741–763

Smith N (2007) Nature as an accumulation strategy. In L Panitch and C Leys (eds) *Socialist Register 2007: Coming to Terms with Nature* (pp 1–36). London: Melrin

Sneddon C (2007) Nature's materiality and the circuitous paths of accumulation: Dispossession of freshwater fisheries in Cambodia. *Antipode* 39(1):167–193

Sterk W and Wittneben B (2006) Enhancing the clean development mechanism through sectoral approaches: Definitions, applications and ways forward. *International Environmental Agreements: Politics, Law and Economics* 6(3):271–287

Swyngedouw E (1999) Modernity and hybridity: Nature, Regeneracionismo, and the production of the Spanish waterscape, 1890–1930. *Annals of the Association of American Geographers* 89(3):443–465

Taiyab N (2006) Exploring the market for "development carbon" through the voluntary and retail markets. *Markets for Environmental Services Series*, Issue 8. London: International Institute for Environment and Development

TWP (2010) *Trees, Water and People: Forest-Saving Stoves Emissions Testing.* http://www.treeswaterpeople.org/stoves/info/emissions_testing.htm (last accessed 14 May 2010)

UNEP-Risoe (2010) *UNEP Risoe CDM/JI Pipeline Analysis and Database.* http://cdmpipeline.org/ (last accessed 8 February 2010)

UNFCCC (2001) Decision 17/CP.7—Modalities and procedures for a clean development mechanism as defined in Article 12 of the Kyoto Protocol. http://cdm.unfccc.int/EB/rules/modproced.html (last accessed 2 June 2009)

UNFCCC (2005) *La Esperanza Hydroelectric Project Honduras Project Design Document.* Bonn: UNFCCC

Vandenbergh M P and Ackerly B A (2008) Climate change: The equity problem. *Virginia Environmental Law Journal* 26:55

Wara M W (2007) Is the global carbon market working? *Nature* 445(8):595–596

Wara M W and Victor D G (2008) A realistic policy on international carbon offsets. *Program on Energy and Sustainable Development Working Paper* 74:1–24

World Bank (2006a) *CDM and JI Methodology: Status Report on Progress and Lessons Learned.* Washington DC: World Bank

World Bank (2006b) Honduras hydropower project makes history today for the small community of La Esperanza. http://web.worldbank.org/WBSITE/EXTERNAL/COUNTRIES/LACEXT/HONDURASEXTN/0,contentMDK:20690511~menuPK:295076~pagePK:141137~piPK:141127~theSitePK:295071,00.html (last accessed 17 January 2007)

World Bank (2008) *CDCF Monitoring Report: La Esperanza.* Washington DC: World Bank Community Development Carbon Fund

Yamin F (2005) The international rules on the Kyoto mechanisms. In F Yamin (ed) *Climate Change and Carbon Markets* (pp 1–74). London: Earthscan

Chapter 3
Making Markets Out of Thin Air: A Case of Capital Involution[1]

María Gutiérrez

Introduction

In the spring of 1997 several indigenous Tzeltal farmers in Chiapas, Mexico, received the first payment for what was a 3-year agreement to offset the emissions of carbon dioxide from Formula One motor sport. The farmers' pledge was to develop and preserve a sustainable forest and agricultural system that would absorb the 5500 tons of carbon emitted every year by the racing cars. In the process, significant areas of cloud forest would be preserved, and with them several rare species such as the resplendent quetzal and ocelots.

This was one of the earliest examples of what is now broadly accepted as a practical if partial solution to both the global increase in greenhouse gas emissions that lead to climate change and to seemingly intractable deforestation. At the time, the hope was that offsets from carbon sequestration by trees—so-called "sinks"—sold worldwide to companies that needed to reduce their greenhouse gas emissions would become "a new crop for farmers in the tropics".[2]

Seven years later, when delegates negotiating the Kyoto Protocol to the UN Framework Convention on Climate Change (UNFCCC) finally agreed on the rules and procedures to govern the new market in emission reduction credits from forestry, it was clear that they had created a scheme that even they could hardly understand. With more than 40 separate decisions and numerous fixes and exceptions, the rules on sinks are widely agreed to be convoluted and disjointed, fully intelligible only to a small number of experts in the world (Fry 2002, 2007; Yamin

The New Carbon Economy, First Edition. Edited by Peter Newell, Max Boykoff and Emily Boyd.
© 2012 María Gutiérrez. Book compilation © 2012 Editorial Board of Antipode and Blackwell Publishing Ltd.

and Depledge 2004). It also became clear that under the regulations applying to the new market in developing countries, most of the existing projects, including the aforementioned Chiapas project, would not be able to compete. Given the long-term and non-permanent nature of carbon sequestration by trees, projects have to be implemented for at least 20 years and comply with many complex, time-consuming and technically expensive steps to ensure accountability and credibility—procedures that are only affordable to large operations. The involvement of small-scale peasant farmers is now considered too costly and too risky and has, at best, become a "niche" market. This is so despite the fact that it is generally small-scale peasant farmers who occupy the agriculturally marginal land that is most appropriate for and in need of forest conservation and restoration, and who have the least resources to dedicate to this. To address this bias, a special category of small-scale projects implemented by low-income communities was introduced as a modality. Capitalizing on this new market "niche", the World Bank inaugurated a Community Development Carbon Fund to help small projects with the transaction costs. Their motto was "Carbon with a human face".

This article recounts how this strange idea came to be and what some of its implications are. It describes the problems encountered by UN negotiators as they attempted to abstract, isolate, quantify and commodify a process equivalent to breathing by trees. To understand this new market—made literally "out of thin air"—I first briefly recount how sinks came to be included in the carbon trade under the Protocol and thus provide the historical and political background that helps to understand them. I then outline some of the key issues that had to be sorted out—from a global definition of forest, to separating the direct human effects on sequestration by trees from indirect and natural ones. The way that the resulting system excludes small-scale producers from the market and rewards mostly the few international agencies that validate the projects draws attention to time and risk allocation as important factors in the way that capitalism reproduces uneven development. The last section addresses the question of how this might be seen as an instance of capital involution and compares sinks to the trade in debt and futures that plagues the contemporary financial system and led to the latest economic crisis, putting so many in jeopardy.

For its complexity and deep contradictions that all the while derive from a basic and simple idea—to abstract and commodify everything you can—I suggest that the market for carbon sinks can be seen as a variety of involution, as understood by Goldenweiser (1936), described by Yengoyan (2001:xi), and used by Geertz (1963) for explaining the sawah system of agriculture in nineteenth-century Indonesia, and noted by Katz (1998) in the specific case of capital and the production of nature. Following their lead, I take involution to refer to instances in which the repetition of a pattern becomes dominant and internally more and more complex, but instead of unfolding and evolving into something different, it in-folds and becomes increasingly pervasive and convoluted. Although the process leads to continuous elaboration, the escalating complexity is not transformative but instead inhibits the development of new structures. Insofar as the new market for sinks reproduces uneven development, it results in involution and is not socially transformative.

I see involution in two senses. In the sense pointed out by Katz (1998:46), it relates to capital's requirement of something "outside of itself"—in this case, pre-existing (Harvey 2003:141)—to ensure continuous accumulation, expanded reproduction and stability. Only that this "outside" is found by reworking the inside. But involution applies also to the entanglement that resulted from commodifying something like carbon sequestration from trees and vegetation, a process which takes place naturally everywhere, anyway. The "scientific virtuosity" and "technical hairsplitting" (Goldenweiser 1936:103) required to turn this process into a fungible commodity has created a very small number of experts in the world and a highly specialized field, and may at best result in improvements in forest and land management in limited places, and at worst in increased competition for productive land and further impoverishment of peasant farmers. Moreover, it distracts from the fundamental concern of reducing the greenhouse gas emissions that lead to climate change—the bulk of which are the result of fossil fuel use.

With negotiations underway on a more comprehensive agreement on climate change, many aspects of the regime have been placed in doubt (including the Kyoto Protocol as we know it)—but not the market for carbon offsets. Some proposals have been made to increase access to the carbon market by developing countries and low-income communities, yet the key challenges remain. At the same time, new initiatives to reduce emissions from deforestation and forest degradation in developing countries (known as REDD+) are also receiving much attention and resources. It is in light of all this ferment that a review of the history of this invention is provided here, as part of an attempt to better understand its potential and implications.

"From the point of view of the atmosphere"?

Carbon sequestration appeared as a simple idea in principle. Trees and vegetation absorb carbon from the atmosphere through the process of photosynthesis. This carbon can be measured, albeit not with great accuracy. With deforestation, the carbon is released back into the atmosphere, contributing to the greenhouse effect—as well as to loss of water retention by the soil and other processes often leading to environmental degradation. But because the problem of climate change was early on conceived *"from the point of view of the atmosphere"*, it did not matter where in the world a reduction in emissions took place.[3] Thus trees and vegetative growth in one place could offset carbon emitted anywhere else. It also happens that areas primarily suitable for forests in developing countries tend to be populated by the poorest peasant farmers, because they are isolated and often located on steep hillsides and are difficult to work, and because the soil cannot productively sustain intensive agriculture without a great amount of agricultural inputs; attempts to extract something out of this land usually lead to increased land degradation and poverty. So even though it was always recognized that carbon sequestration was just a temporary solution of limited effect, investing in carbon sinks appeared as a simple and inexpensive solution that addressed both the increase in greenhouse gas emissions and deforestation—two major environmental problems—at the same time.

Yet once the idea of accounting for carbon sinks in the context of a market was accepted, an interminable list of problems appeared. How does one define a forest in a way that applies to all countries and contexts? Even the common definition of forests has changed in the last few years, having evolved from notions that included the idea of climax, to references to ecosystem, and most recently images of an "erratic, shifting mosaic of trees". But then what about savannas and woodlands? And short shrubs, mangroves, or marshland? And forest transitions? What makes the removal "human-induced"? How does one set apart natural regeneration or fertilization from other natural processes or past practices? How does one interpret "since 1990"? Is it fair to use the same baselines for all countries? What happens when the carbon reverts back to the atmosphere as a result of forest fires, pests and other natural disturbances? Will setting aside land as carbon stock result in other areas being affected from the displacement of previous activities? How does one verify and ensure accuracy? Clearly these questions were hardly just technical. In fact, in the context of a new global market any call to distinguish between political and technical issues in regards to sinks was quickly considered politically motivated.

For sinks to function as a commodity they had to be fungible with other carbon offsets. This meant that the carbon taken up by a tree as it grows over time had to be made to correspond to the carbon emissions avoided when closing down a coal plant for good—ton per ton; it required making something gradual and non-permanent equivalent to something immediate and permanent. Because trees cannot guarantee that the carbon taken up will be stored forever—on the contrary: their status as living things guarantees that it will eventually be released—this task was most complicated.[4]

Critically, real offsets in emissions have to be proven to have taken place, because under an international market mechanism, for every credit bought in a developing country an equivalent amount of greenhouse gas will be emitted into the atmosphere in an industrialized one. Failure to really offset leads to doubling emissions instead of halving them. As part of a United Nations international environmental and sustainable development agreement which, it is mandated, should be public and transparent, the market has to be credible and legitimate.

In order to sort out these problems and assure the credibility of the market, delegates at the UN came up with an incredibly complex set of rules for accounting for sinks. The complexity is such that, in regards to the trade with developing countries, transactions costs take up the largest share of the investment, making smaller projects or those with numerous participants nonviable. A brief recount of how this came to be is what follows.

Negotiating Sinks

Together with emissions trading, carbon sinks appeared very early on in the negotiations under the "comprehensive approach" to climate change. This entailed collectively accounting for removals as well as emissions of different greenhouse gases and measuring them according to a single metric—that of their global warming potential (GWP).[5] The main proponent of this approach was the United States.[6] Almost everybody else also approved of the approach in theory; its advantage was that it allowed countries to choose the most cost-effective emission reductions.

Sinks were thus included in the 1992 UN Framework Convention on Climate Change (UNFCCC) as part of countries' overall efforts to mitigate and adapt to climate change.[7] But accounting for sinks as part of quantified mitigation commitments under the Kyoto Protocol was something else, and depending on the definitions and modalities allowed, their effect could be as broad as to make any emissions limitation unnecessary. As the Intergovernmental Panel on Climate Change (IPCC) stated at the time, terrestrial ecosystems sequester globally an average of 2.2 GtC per year through natural regeneration; if countries received credit for even half of these not-human induced, indirect effects of sinks (mainly naturally occurring carbon or nitrogen fertilization), they could meet their commitments without any additional measures (IPCC 2000:80). Yet it was only in the later stages of negotiations to the Protocol that countries realized how much was at stake.

Although calculating removals from sinks is far from accurate, when numbers began being crunched they showed that for most of the

Scandinavian countries as well as the USA, the Russian Federation
and some Eastern European countries, annual domestic sinks accounted
for at least 10% of national gross emissions alone, even without any
direct policy to promote them. In New Zealand, Sweden, Latvia and
possibly Finland, for example, total net sinks could absorb more than
half of their total emissions. The case was very different for most of
Western Europe, where high population densities and intensive land use
patterns limited the potential of domestic sinks to trim down emissions
accounts.

Many developing countries generally saw sinks as yet another
loophole by some industrialized countries to avoid cutting back on
emissions while giving the impression of doing so. But when it came to
include them in the Kyoto Protocol's Clean Development Mechanism
(CDM)—under which emission reduction projects undertaken in
developing countries result in emission reduction credits advanced to
developed countries to comply with their mitigation commitments—
views were more divided. For some countries (notably Brazil) the
inclusion of sinks brought up sovereignty concerns relating to land use.
But for many others it represented the possibility of additional resources
and investment in the forestry sector or in conservation.

A key aspect about sinks to keep in mind is that sinks are mainly
a negotiating tool in the wider context of overall emission reduction
commitments. For the largest emitters and several other countries with
important forestry or extensive agricultural sectors (the USA, Australia,
and Canada, as well as Russia, Norway and New Zealand for example),
including sinks in overall accounting allowed them to assume the
commitment to cut emissions much more easily. For the USA, including
broadly defined sinks in the accounting of the Kyoto Protocol meant that
its reduction target went down from -7% to -3% or -4% relative to 1990
emissions (Grubb 1999:117).[8] Sinks were thus said to be "the ultimate
flexibility mechanism"—and these countries negotiated accordingly.

But sinks were also of interest to many Latin American countries and
some African ones, which saw them as a potential source of income in
a sector where foreign investment is scarce, and where the challenge
is compounded by international pressure for forest conservation. For
some poor African countries, whose emissions from fossil fuel burning
account for less than 2% of the global total, they represented the only
opportunity to participate in the Kyoto market. And precisely because
they meant so much for some of the largest emitters, countries like
the small island states with no reduction commitments but with serious
concerns for the effects of climate change sometimes used sinks as a
bargaining chip to press for more substantive action on other issues. A
representative of the Alliance of Small Island States (AOSIS) referred
informally to sinks as *"the only leverage we have"*.[9]

Yet in many ways this bargaining chip grew out of proportion as people started the work of trying to delimit and make carbon sequestration from land use change quantifiable and verifiable, and there is a widely held sense that very few people had any idea what they were getting into. Simply, some of the technical challenges are insurmountable. Something as fundamental as clearly distinguishing direct human-induced changes from indirect and natural ones cannot be resolved scientifically, and can only go back to the negotiating table to find a political solution.[10]

Matters were aggravated by the fact that under the Kyoto Protocol countries took up specific quantitative reduction commitments without clear knowledge or agreement of the rules that would allow them to meet those commitments. Individual country targets were thus adopted in Kyoto in December 1997, while the detailed rules were left to be agreed in The Hague in November 2000. A process ensued whereby countries attempted to create rules to help them achieve the targets earlier accepted.

Take the situation of the USA: as Grubb and Yamin (2001) explain, in the year 2000, emissions there were at 13 percent above 1990 levels. Assuming strong mitigating action took place (putting aside political challenges of implementation at the national level), emissions would probably be up to 10 to 20% above 1990 by the end of the first commitment period in 2012. The chances of the USA meeting its commitment under the Protocol of minus 7% compared to 1990 levels were slim, requiring not only prompt and radical changes in energy investment, but also emission credits for more than 200 million tons of carbon a year, acquired either through the flexibility mechanisms or through sinks.

In fact, the USA had always assumed full net-net carbon accounting of sinks in its calculations, and had included them when it took up its minus 7 percent commitment. In January 1998 the State Department explained in a fact sheet that 4 percent of the 7 percent reduction would come from "certain changes in the way gases and sinks are calculated," so that the US commitment amounted to "at most" three percent in real reductions (see Schlamadinger and Marland 2000:44).

But the definition and rules for accounting of sinks kept changing as negotiations proceeded, which meant that countries kept having to reposition themselves as they recalculated how their targets and commitments were affected. Granted, agreement by more than 190 countries with completely divergent interests requires some level of ambiguity. Yet this can backfire when the time comes to define operating rules. This is precisely what happened in The Hague, when largely as a result of this process negotiations broke down. When negotiators met again in Bonn, Germany, six months later to finalize their work, the

USA had withdrawn, and in so doing had altered the whole negotiation landscape.

Without the USA it was no longer a matter of ensuring the environmental integrity of the Protocol, but of securing support for its survival. In order to obtain agreement at the resumed session in Bonn in May 2001, Australia, Russia, Japan and Canada were allowed to offset over half of their emissions with carbon sinks under Protocol article 3.4 as part of a package deal.[11] The deal, transformed later into the Marrakesh Accords, also contained limits to sinks in the CDM: for the first commitment period, avoided deforestation was left out of the CDM and only afforestation and reforestation were allowed, with credits from these activities capped.[12] To compensate, the package deal also included a set of principles meant to assure that any reductions claimed for removals from LULUCF are real and additional to any that would have happened anyway and to avoid perverse incentives from LULUCF activities.[13]

Thus political agreement in Bonn on the dividing "crunch issues" was hailed as a big success, even if it had meant halving the reduction commitments originally agreed at Kyoto (see Ott 2001). The allowances on sinks, together with concessions on the tradability of emissions rights, led some commentators to conclude that Kyoto had been reduced to mere symbolic policy of little environmental effectiveness, codifying more or less business-as-usual emissions and making it cheap to comply with any commitments (Böhringer and Vogt 2004).

The Problem with Sinks

Because sinks were to work as a quantifiable and tradable unit, and eventually as a commodity, they had to be very clearly defined. This meant isolating, carving up, and putting a figure on the highly complex organic process of terrestrial carbon exchange and the even more complex social one of land use change. In the IPCC Special Report prepared to address sinks there is hardly a page that does not mention the word "uncertainty". In the end, most of the problems were somehow resolved through a combination of definition and accounting fixes. Here I elaborate on two of the most basic issues to be sorted out: the definition of forests, and the problem of separating direct human-induced effects from indirect effects and natural disturbances.

A Global Definition for Forests

The first question that had to be settled was the definition of forests and the related meaning of afforestation, reforestation and deforestation. As part of a negotiated UN agreement, the definitions have to be applicable to all parties and be simple, based on accessible data, and subject to verifiable accounting. They also have to be consistent to allow for their inclusion in the carbon market under the Kyoto Protocol. Clearly, this was no easy task. The IPCC Special Report on LULUCF cites an article listing 240 definitions of forests used by countries, depending on their social and economic structures and biogeophysical conditions—and these are just some official ones (Lund 1999, in IPCC 2000:63). These definitions have for the most part little to do with carbon content.

Negotiations on this issue began revolving around a general definition of forest used by the Food and Agriculture Organization (FAO) based on land cover, which establishes a threshold of minimum canopy cover—that is, the proportion of ground area covered by tree crowns. But again, this is not a widely applicable or accepted definition. The IPCC reported that, globally, about 50% of wooded land has a canopy cover of less than 20%, and can vary at the national level between 10% and 70% (IPCC 2000:64). Variations between regions and between countries are also great. One can have a closed canopy moist forest in one place, and a sparsely treed, low canopy cover savanna in another.

In pure carbon accounting terms, establishing a precise and global definition simply based on canopy cover is highly problematic. If the threshold is low, say a 10% canopy cover, a dense forest could be seriously degraded and result in high carbon emissions without qualifying as forest loss or deforestation. If the threshold is high, for example 70%, significant areas could be cleared without the resulting loss of carbon being accounted for. Conversely, any increases in the canopy cover beyond 10% (up to 90%) would not comply with the definition of reforestation or afforestation and would not accrue credits for carbon sequestration.

There is also the problem of measuring actual canopy cover without consideration of potential canopy cover. Since definitions under the Protocol are about a change in land use—from forest to non-forest in the case of deforestation for example—a definition of forest based strictly on actual canopy cover could lead to harvesting falling under deforestation, or to natural regeneration being referred to as reforestation. Similarly, if potential canopy cover at maturity under planned land-use practices was the basis of the definition, carbon loss from harvesting or carbon sequestered from regeneration activities might not be counted (IPCC 2000:6).

This points to the problem of timing, that is, the great asymmetry in the rates at which carbon is released into the atmosphere and the rate at which it is recaptured as the forest grows again—decades and even centuries, depending on the species of trees and site conditions. Yet the carbon uptake needs to be tallied by a specific date to count towards fulfillment of commitments under the Protocol and to function in an active market. A regenerating boreal forest stand, which could require decades for its canopy cover to reach the definitional threshold, would reach this maturity level well beyond the 5-year commitment period established by the Kyoto Protocol.

After many hours of negotiating time, forests were defined as follows:

> "Forest" is a minimum area of land of 0.05–1.0 hectare with tree crown cover (or equivalent stocking level) of more than 10–30 per cent with trees with the potential to reach a minimum height of 2–5 metres at maturity *in situ*. A forest may consist either of closed forest formations where trees of various storeys and undergrowth cover a high proportion of the ground or open forest. Young natural stands and all plantations which have yet to reach a crown density of 10–30 per cent or tree height of 2–5 metres are included under forest, as are areas normally forming part of the forest area which are temporarily unstocked as a result of human intervention such as harvesting or natural causes, but which are expected to revert to forest.[14]

This definition of forest, contrary to common perceptions of what a forest is, contains no mention of biodiversity, for example. Nor is there reference to sustainable use of natural resources, which one would expect of an international environmental treaty dealing with forestry activities. More pointedly, it does not distinguish between "natural" forests and plantations. This means that an oil palm plantation could appear as forest and get credit for afforestation or reforestation. Many environmental NGOs and others expressed concern that in order to cash in on reforestation credits natural forests could be cut down to install fast-growing plantations, and campaigned strongly against sinks for this reason. Their concern was partly addressed with the definitions of afforestation and reforestation, which require that there be no forest on the land for the past 50 years (afforestation), or since 31 December 1989 (reforestation).[15] Thus in defining forests as being either natural or a plantation, cutting down a forest to establish a plantation would not accrue credits from reforestation, as they are officially both "forests"— yet, for the same reason, doing so might also not be considered as deforestation, even though it is.

There is clearly a difference between a multi-species forest ecosystem and 10,000 hectares of fast growing eucalyptus, pine or gmelina, genetically modified to produce higher yields of more uniform woods. To equate these two has serious policy implications, and the lack of distinction in the words could serve to promote monoculture plantations as an effort to counter deforestation or advance reforestation by "increasing forest cover". Furthermore, big monoculture plantations often have negative impacts on the soil, water, plants and wildlife, they create very few jobs, mostly low quality, and do not generate wealth at the local level—canceling out more options for the local people than creating them.[16]

Direct Versus Indirect Effects and Natural Disturbances

In order to avoid crediting business-as-usual, the Protocol makes clear that for land use activities to count as carbon sinks and generate credits, they have to be the result of a "direct human-induced" action. However, distinguishing what is a natural or an indirect effect from a human-induced one is far from easy. Even the most common example of a natural cause of sink reversal, forest fires, is tricky. These can be the result of a natural event such as lightning, or of a direct or indirect human act, such as accidental fire, arson or prescribed burning and its escape. The causes are in many cases not simple to attribute. Much more complex is discounting for indirect nitrogen deposition, elevated carbon dioxide concentrations above their pre-industrial level (ie current greenhouse gases accumulated in the atmosphere), and activities and practices before the reference year.[17]

In the case of fire, a definition that assumes forest fires as carbon loss might not even be consistent with the long-term maintenance of forests or the increase of carbon stocks. Fire is in fact a natural part of many forest ecosystems, as well as a forest management tool. Natural or prescribed, fires work as breaks that reduce the chances of more intense fires spreading. So even though they may be beneficial and necessary, in the short time of a commitment period they appear simply as carbon loss. In such cases, carbon accounting could serve as a disincentive to the sustainable management of the forest. To complicate things further, it appears that the fire regime in parts of the world is changing, due possibly to direct and deliberate human manipulation, as well as indirect manipulation—including climate change. Fire regimes will thus vary with rapid climate change or in response to El Niño pattern changes. Furthermore, forest fires do not always result in complete tree mortality or in deforestation as commonly defined. In eucalyptus and some pine forests, most mature trees will survive even intense fires. Most trunks and large branches will regrow full canopies in a few years—but probably not in time for accounting in the commitment period.

Addressing these so-called "natural disturbances" has in fact become a key issue in the negotiations on LULUCF rules post 2012, and a few countries (in particular Canada and Australia) have provided vivid examples of some of the problems with commodifying sinks and using them to meet emission reduction commitments. For instance, in 2003 wildfires in south-eastern Australia resulted in emissions of 190 million tonnes of carbon dioxide equivalent ($MtCO_2e$) from existing forest lands. Australia has annual allowable emissions during the first commitment period of 591.5 MtCO$_2$e. This means that a third of its whole economy emissions in one year was simply taken up by wildfires—hardly the free ride that countries at one point thought sinks could be.

Sinks under the CDM

Besides the above-mentioned problems, sinks in the CDM pose an extra set of challenges. Because developing countries have no quantitative mitigation commitments, any emission reductions achieved within a project may be cancelled out by emissions outside the project with no accountability or concern. As such, the market offer is potentially unlimited, and both seller and buyer have an incentive to inflate the emission reductions achieved.[18] To address the risk of "flooding" the market with cheap credits, the rules limit eligible sink activities under the CDM to afforestation and reforestation in the first commitment period, and establish a quantitative cap of 1% of a buyer country's baseline emissions for each of the five years of the commitment period. However, at the project level, problems intrinsic to sinks remain. These include the following.

Non-permanence

A biological carbon sink today may be a source of carbon dioxide tomorrow, whether for natural or human-induced reasons, including climate change itself. This is referred to as the problem of non-permanence, and any accounting of emission reductions from land use change would have to consider it in light of the need to make all credits fungible under the market—whether permanent or non-permanent. A major quandary is the long time required for carbon sequestration by trees, and the short time in which the carbon sequestered may be lost back to the atmosphere.

To address this problem, parties agreed to a temporary approach by which the carbon credits obtained by a developed country from forestry projects under the CDM expire when the carbon is emitted back to the atmosphere for whatever reason. Then, the holding country either has to reduce national emissions by that amount or buy the same number of carbon credits from another forestry project. Two new types of certified emission reductions (CERs)[19] were created to account for sinks: temporary CERs (tCERs) and long-term CERs (lCERs—also known as ul-CERs). The crediting period was established at either 20 years—which may be renewed at most twice (up to 60 years)—or a maximum of 30 years.[20]

Still, trying to solve the problem of non-permanence solely by accounting can bring up other problems. For example, liability gets pushed forward in time. NGOs and some Parties feared that a situation could arise in which, once the lifetime of the tCERs comes to an end and replacements are needed, there could be a spike in demand for emission reductions, which would weaken further commitments, and which might ultimately mean that in fact the promised and additional permanent cuts would never materialize. To ensure that parties do not accumulate too much debit for the future and to guarantee that credits are replaced in a timely way, each buyer country has to include in its national registry lCERs and tCERs replacement accounts for each commitment period.

This treatment of non-permanence through temporary crediting is a major factor affecting the competitiveness of sinks compared with other credits under the CDM: not only is the project cycle long and the requirements more complicated, but even with these hurdles overcome the credits have to eventually be replaced and are therefore not exactly fungible, making their price lower. Only the largest operations with economies of scale can hope to be competitive in such conditions.

Additionality

Another key challenge for emission reduction projects under the Protocol's carbon market is proving that the reductions would not have taken place in a "business as usual" situation—that is, in a world without credits. This is known as the problem of "additionality". Because the carbon market was originally conceived as a first step to induce the development and transfer of greener technologies, the idea was that all certified reductions awarded must not have happened anyway but be additional.[21] This sounds more obvious than it is in practice, and proving additionality became one of the most difficult and controversial issues under the CDM.

Because forests and other biomass tend to grow back naturally if left undisturbed, in the case of sinks additionality would have to be calculated in excess of any natural regrowth, unless it was proven that the land would be used for other purposes where this natural regrowth would be artificially prevented. But there are also technical problems to consider. In some wet areas, for example, an increase in plant biomass carbon can lead to a decrease in soil organic carbon, possibly offsetting any gains in sequestration made by the plants. This has been the case in certain grasslands of the southwestern USA, where woody plant invasions of wetter grassland resulted in a loss of soil organic carbon (Jackson et al 2002). A similar problem has been pointed out for intensively managed tree plantations which, once their productive life is over, tend to diminish the natural regrowth capability of the soil.

The final decision on additionality parallels that for energy projects in that it allows the Executive Board of the CDM to decide on a case by case basis which projects are additional.[22] Yet this leaves the question open to interpretation and the issue remains controversial, with many parties and non-parties calling for its reconsideration.

Leakage

Because projects do not operate in a vacuum, a reduction of emissions in one place may result in an increase elsewhere, as when land is cleared for a plantation that generates no employment and people are displaced, often clearing forest land in another place, or when demand for timber, fuelwood or other goods is simply relocated. This "externality" is referred to as "leakage". Commercial plantations are particularly prone to leakage, which is another reason why many environmental groups oppose them.

This is a matter of consequence for the ecological and social integrity of the Protocol's mechanisms. For example, a study in the USA of a large carbon sequestration program to convert agricultural land to tree plantations found that the benefits from the offsets would eventually be lost, as landowners responded by harvesting existing forests and converting unsubsidized land back to agriculture (Alig, Adams and McCarl 1998). Under a market for credits, it is not enough to acknowledge leakage—it should be quantified and reported.

The final decision defines leakage rather narrowly as "measurable and attributable to the CDM project activity", and merely states that projects must account for it by including a monitoring plan in the project design document that provides for procedures for the periodic calculation of reductions and for leakage effects, and by adjusting for leakage in the calculation of credits.[23] Still, some leakage—including international leakage, where emissions are displaced into other countries—may not easily be attributed to the original project activity.[24]

Baselines

Any accounting of carbon sequestered by a specific project needs a reference or base from which to start counting—that is, a baseline above which carbon uptake has increased as a result of the project. A project proposal therefore implies two things: developing a reference scenario for future human activity on the site, and estimating the carbon stock under this scenario relative to the base. In the case of land involving many people, making such assumptions is difficult at best, since neither human nor natural activity is ever static; they respond to a multitude of stimuli from various scales. In the case of conversion of agricultural or cattle grazing land to plantations one may assume that, all things being equal, existing human activity and carbon stocks would persist—the problem is of course assuming this "all things being equal" not for one or two years, but for the duration of the project, which can go from a couple of decades to 60 years and more.

To address this problem, projects develop a scenario using site-specific information such as plans and inclination of the landowners, as well as regional and national economic trends and policies. But extrapolating data from regional and national trends and then forecasting for a land use project, even while taking into account its specific particularities, is merely a speculative exercise and is always uncertain. The causes behind farmers changing the land use may have to do with poor agricultural practices in situ or with apparently unrelated government policies which in many cases are linked to actions and events well beyond national borders.

There is nothing much negotiators can do in establishing general rules for this. The decision therefore states that baseline methodologies are to be developed based on existing or historical changes in the carbon stocks, those expected from a land use that represents an economically attractive course of action, or from the most likely land use at the time the project starts. Moreover, in order to play it safe and provide a conservative account of the baseline, and to prevent earning credits for avoided emissions that result only from the displacement of the previous land use, the final decision states that greenhouse gas emissions from activities on the land before the project was implemented are not to be included in the baseline. In the case of conversion from cattle grazing to plantations for example, the emissions from the cattle that would occur had the cattle stayed and that are avoided by switching to trees, are not be counted in the baseline.

Social and Environmental Impacts

The social and environmental impacts of sink projects under the CDM were always an issue and it was one of the few instances where negotiations turned to actual effects of the projects on people on the ground. Tuvalu, on behalf of AOSIS, and the European Union with Norway and Switzerland, had each introduced a proposal detailing a list of topics that had to be covered in the analysis of environmental and socio-economic impacts of project activities. But these were strongly opposed by most developing and some developed countries generally on the grounds that the definition of a standard list of sustainable development criteria would impinge on national sovereignty. Many of these countries also opposed it on more practical grounds, noting that such an assessment would significantly augment the project transaction costs—which were already anticipated to be very high and which the project developers in the host countries would have to bear.

So when it came to the decision, national sovereignty concerns and transaction costs ruled: in the application of CERs, only a general list of areas of enquiry for environmental and socio-economic impact assessment are required.[25] Besides conditions of the area (such as the possible presence of rare or endangered species and their habitats, or the current land tenure and land use rights of access to the sequestered carbon), the list includes possible impacts on biodiversity, natural ecosystems and impacts outside the project boundary, as well as possibly information on local communities, indigenous peoples, local employment, food production, cultural and religious sites, access to fuelwood and other forest products. If any negative impact is considered significant by the project participants or the host party, the rules further require a statement that project participants have undertaken an environmental and a socio-economic impact assessment, in accordance with the procedures required by the host party, and a description of planned monitoring and remedial measures to address such significant impacts. The designated operational entity in charge of validating the project is only to verify whether this information has been submitted, but is not to assess it. Stakeholder comments on the project need merely be "invited" by project developers. As the Climate Action Network said, "Let's hope they're home to get the invitation and not working on subsistence agriculture in fields that might soon be displaced by *Eucalyptus.*"[26]

In its final form, then, the text is open enough for all to agree and there are no real requirements of consequence to address social and environmental impacts of the projects. It is all in accordance with host Parties' internal procedures, and no follow up is required. Whether local people's interests are taken into account will depend on each specific project and on the amount of publicity and pressure civil society groups may bring to bear on buyers and host countries.

On the Role of Time and Risk in Reproducing Uneven Development

The new market in emission reductions can only be conceived in the context of uneven development. Its logic and justification make sense solely in a situation of geographical and politico-economic difference, and the Kyoto Protocol's flexibility mechanisms are designed precisely to take advantage of this difference (Harvey 2006). In the case of the market for sinks, the high transaction costs derived from complying with the rules described above contribute to the exclusion of most forest producers except the well capitalized and those who already have land and time to spare, thereby further reproducing uneven development.

Two factors are salient in this reproduction of uneven development: time and risk. They apply to both the two commodities analyzed here, that is, trees and emission reduction credits. In the case of trees, the time necessary to capitalize on a plantation becomes a factor that by itself tends to exclude small landowners and poor peasant farmers, even if the land they hold is most suitable for this kind of activity. Because trees take years to grow, land gets locked-up in waiting for the first thinning and the first cut. Small landowners hardly have the luxury of leaving their land unproductive for such long periods. It is normally medium-size and big companies with access to credit and large tracts of land and time (allowed by additional or parallel sources of income) that can afford to establish these types of plantations.

But analyzing access to the market for emission reduction credits also highlights time as a discriminating factor in the sense that the market for sinks, as fictitious capital, is highly dependent on the creation of credibility and financial credit (Harvey 1999 [1982], 2001). Plotted as a *filière* or commodity chain (see Ribot 1998), it is the time that has to be allotted for each activity and the rate of return on that investment that count. There is a big difference between the time devoted by project developers to the preparation for and implementation of a project, and the time spent by consulting agencies in validating or certifying the project vis-à-vis eventual proceeds. Project developers in developing countries need substantial amounts of upfront resources to cover the project until the end or until they can sell the emission reduction credits. In contrast, established certifying agencies (most of them from developed countries) will cash in on their consulting services almost immediately, freeing time and resources for further investment. In cases of fictitious capital like the one analyzed here, credit, and credibility, become key resources. And both are tightly linked to time.

Risk is the other factor that clearly affects who has access to the market for sinks and how profit is made. In the case of the CDM, risk is a central feature of the mechanism and a key discriminator affecting the price of credits. The risk of failure is more insidious in sink projects given their nature, the length of time that trees need to grow and the complexity of the rules that govern them. Most poor peasant farmers cannot afford to take the level of risk required; they lack the back-up infrastructure or reserve resources to draw on in case of failure. Focusing on how the burden of risk is shared is essential to understand markets and power, and determining who bears the brunt of risk in any particular project illuminates how inequality may be reproduced.

On Capital Involution

Geertz used the term "agricultural involution" to refer to "the ossification of the Indonesian agrarian economy" in the twentieth century (Geertz 1963:38). Key to this ossification, he argued, was the sawah system of wet-rice agriculture, with its ability to support a rising demand through intensification, increasing technical complexity and specialization. He noted: "the output of most terraces can be almost indefinitely increased by more careful, fine-comb cultivation techniques; it seems almost always possible somehow to squeeze just a little more out of even a mediocre sawah by working just a little bit harder" (Geertz 1963:35). Thus, "[e]ven the most intense population pressure does not lead to a breakdown of the system on the physical side (though it may lead to extreme impoverishment on the human side)" (1963:33). At a more general level, "[...] the mere quantity of preparatory (and thus not immediately productive) labor in creating new works and bringing them up to the level of existing ones tends to discourage a rapid expansion of terraced areas in favor of fragmentation and more intensive working of existing ones" (1963:36).

Geertz argued that this ability of the sawah system allowed it to absorb an increased population and provide part-time labor force to work the Dutch-run sugar plantations, mitigating the impact of capital-intensive industrial agriculture in Indonesia in the second half of the nineteenth century. It provided "one-foot-in-the-terrace, one-in-the-mill" labor that was most convenient for capital accumulation (siphoned off by the colonial managers) without the attendant structural changes in the indigenous economy. This resulted over time in a stifled agrarian sector, handicapped by an increasingly severe dual economy, split between a mass of impoverished peasants and a small managerial elite, and unable to profit from advanced technologies and modern institutions.

I argue that this "concentrative, inflatable quality" (Geertz 1963:37) of a sawah can be seen in nature as an accumulation strategy for capital when looking at the market for sinks. Making sinks a commodity has required an extraordinary amount of abstracting and oversimplifying, but also of accounting fiddling that has resulted in exceptional complexity. This Byzantine complication results mainly from the development of a simple idea or pattern within capitalism: to abstract, single out, privatize, assign value, account and trade anything you can think of. "The inevitable result is progressive complication, variety within uniformity, virtuosity within monotony. This is *involution*" (Goldenweiser 1936:103).

This concept of involution is not proposed here as an alternative to Marx's account of the creation of fictitious capital and commodification under capitalism—quite the contrary: it confirms Marx's points. Rather, it is suggested as a metaphor to illustrate how capital manages to extract ever greater profit from nature while failing to deliver substantial change, and in the process generates increased entanglement. It speaks to the "involuted" quality of the commodification of nature pervading the process of capital accumulation today: the system grows ever more intricate, relationships more complicated, arrangements more complex—all in order to extract ever more profit which benefits ever fewer people.

Recent efforts to gain more from both money and nature are in many ways similar: they consist of further breaking down and carving pieces that can be sold at a profit, and reshuffling them to spread risk. The complex mortgage-backed assets engineering that led to the current financial crisis can thus be seen as another example of capital involution. But the similarities go beyond the complexity created.[27] Insofar as the market for sinks—and offsets in general—consists of the promise to do something—or rather, not to do something—it represents a futures contract, "an agreement made by its buyer to take delivery of a specific commodity on a specific date, and by its seller to make delivery" (Henwood 1997:29). This market also signifies a trade in debt: it is premised on the expected demand by companies with emissions debt—that is, those companies that in emitting more than their share have "borrowed" allowances. Like the financial system's wide spread of debt that characterizes the capitalist economy of late, the market for carbon offsets could be seen as an extension of the creation of debt from the economic system to the ecological system. In the same manner, the debt generated is passed on to the public—in this case, when in default, resulting in intensified climate change affecting generations to come everywhere, but particularly those most vulnerable. Risk is so spread out that no one seems to be liable. Like the sawah system, the market for sinks serves to cushion important forces that should lead to substantial transformation—as is the need for a radical restructuring of the energy sector worldwide. The requirement to reduce emissions, which would necessitate significant changes in the economy, is hedged. Yet as even Tim Geithner, US Secretary of the Treasury under President Obama admitted, "the changes that have reduced the vulnerability of the system to smaller shocks may have increased the severity of the larger ones" (in Foster and Magdoff 2009:62).

While it is important not to take this idea of involution strictly, since it is too naturalistic and appears to preclude change, it does seem to reflect this instance of the production of nature under capitalism (Harvey 1996; Smith 2008 [1984]). As exemplified by the case of a market for sinks, the repetition of a discrete, simplistic pattern in confined space results in increased density and entrapment. Instead of opening up options, it generates new problems. Of course every new constraint opens up possibilities in a dialectical process. But as long as the approach taken is narrow—as the reduction of complex social and ecological processes to carbon accounting under a market mechanism clearly is—the real cause of the problem will be left unaddressed, while all efforts go to tinkering with the convoluted arrangements. It is yet another instance of "the myopic pragmatic optimism which allows short-run gains to obscure the general trend of events, which isolates purely technical improvements from the historically created cultural, social, and psychological context in which they are set, and which, because of these failings, exacerbates the ailments it sets out to cure" (Geertz 1963:147).

Endnotes

[1] See Gutiérrez (2007) for a more detailed elaboration of the arguments contained in this article.

[2] United States Initiative on Joint Implementation (USIJI), Klinki Forestry Project, 4 June 1998 (on author's files).

[3] This is one of the most commonly heard phrases in the negotiations, and many of the problems described below stem from this starting point of "the atmosphere's point of view". This idea is linked to Descartes's "mind in a vat" as described by Latour (1999), and is part and parcel of ecological modernization (Hajer 1995; Harvey 1996). The convenient distancing that it affords allows for equating all emissions from all countries, whether they result from fulfilling basic human needs or from unnecessary luxuries. It is likewise the rationale behind accounting for sinks and emissions from land use and forestry as if they were the same as cutting emissions at the source by phasing out fossil fuels.

[4] Sink is defined by the IPCC as "any process, activity or mechanism which removes a greenhouse gas, an aerosol or a precursor of greenhouse gas from the atmosphere". In this sense, sink is more of a verb (a process) than a noun.

[5] Global Warming Potential (GWP) is an index used to compare, by way of a common measure, the radiative forcing of various gases relative to carbon dioxide, using a specific time horizon (100 years in the case of the Kyoto Protocol).

[6] Even before negotiations started for the UNFCCC, the US government had included sinks as "negative emissions" in "Concept Papers" prepared for the IPCC Response Strategy Working Group in December 1989. US Concept Paper: Comprehensive greenhouse gas approach to addressing climate change. 29 December 1989 (unpublished). See Bodansky (1993:517).

[7] The main obligation on sinks stems from UNFCCC Article 4.1(b), which calls all parties "to formulate, implement, publish and regularly update national and, where appropriate, regional programmes containing measures to mitigate climate change by addressing emissions by sources and removals by sinks of all greenhouse gases not controlled by the Montreal Protocol, and measures to facilitate adequate adaptation to climate change".

[8] Under the Kyoto Protocol, industrialized countries committed to reduce their emissions in the first commitment period (2008–2012) to a percentage below 1990 emissions. While these reductions average to around 5.2%, each country committed to a specific number. These numbers are included in Annex B of the Protocol.

[9] On file with author.

[10] For a thoughtful discussion of the precariousness and compromise implied in measuring nature, see Robertson (2006).

[11] Conference of the Parties 7: Land use, land use change and forestry. Decision 11/CP.7 (later decision 16/CMP.1) FCCC/CP/2001/13/Add.1. Marrakesh, 2001.

[12] For a detailed account of the exclusion of avoided deforestation, see Boyd (2003).

[13] The set of principles was proposed by Brazil and the group of developing countries, and initially opposed by many industrialized ones. They state that LULUCF activities should be based on sound science and contribute to biodiversity conservation and sustainable use of natural resources; that their accounting be consistent over time, with any reversal of removals being accounted for at the appropriate moment, and that their

accounting not imply any transfer of commitments to the future; that the mere presence of carbon stocks be excluded from accounting, as well as removals resulting from high carbon dioxide concentrations, indirect nitrogen deposition and other natural effects of past practices on the land; and that their accounting not change the goal of the Protocol to reduce overall emissions by at least 5% below 1990 levels in 2008–2012. See Decision 16/CMP.1.

[14] Annex to decision 16/CMP.1: Definitions, modalities, rules and guidelines relating to land use, land-use change and forestry activities under the Kyoto Protocol. Paragraph 1(a).

[15] According to Decision 16/CMP.1, "'Afforestation' is the direct human-induced conversion of land that has not been forested for a period of at least 50 years to forested land through plating, seeding and/or the human-induced promotion of natural seed sources;" and "'Reforestation' is the human-induced conversion of non–forested land to forested land through plating, seeding and/or the human-induced promotion of natural seed sources, on land that was forested but that has been converted to non-forested land. For the first commitment period, reforestation activities will be limited to reforestation occurring on those lands that did not contain forest on 31 December 1989."

[16] This has been the main argument for those environmental NGOs opposing sinks. They point out that it is not only that plantations cannot substitute forests, having such contrasting environmental impacts, but plantations usually mean the cancellation of the possibility of natural forest regeneration. More than "green deserts", plantations are often referred to as "green wastelands" because "there is more biodiversity in a few square meters of the Namib desert than in an entire plantation" (Carrere 2000: 2).

[17] The IPCC tried to address this problem (known as "factoring out") by distinguishing between "managed" and "unmanaged" forest land—using "managed land" as a proxy for direct human-induced. But this approach remains highly controversial. See Fry (2007).

[18] This affects all projects under the CDM. See Schapiro (2010) for a recent journalistic exposé. For a thorough overview of problems with the carbon market in general, see Newell and Paterson (2010).

[19] Certified emission reductions (CERs) are the credits issued under the CDM.

[20] See Decision 5/CMP.1: Modalities and procedures for afforestation and reforestation project activities under the clean development mechanism in the first commitment period of the Kyoto Protocol.

[21] See Protocol Article 12, paragraph 5(c).

[22] Decision 5/CMP.1 (note 18 above).

[23] Decision 5/CMP.1 (note 18 above).

[24] This is a problem that may be particularly insidious in the case of REDD+.

[25] Decision 5/CMP.1 (note 18 above).

[26] CAN, 7 December 2003. "Chair's proposal for sinks: 'Sinks—close to a decision???'" (copy in files of author). CAN is an international network of over 450 NGOs working to promote sustainable climate change mitigation and adaptation policies.

[27] On a basic level, financial speculation, like the market for sinks, does not lead to sufficient employment relative to the billions of dollars borrowed and speculated with (see Foster and Magdoff 2009), and they both depend on regulation.

References

Alig R, Adams D and McCarl B (1998) Ecological and economic impacts of forest policies: Interactions across forestry and agriculture. *Ecological Economics* 27:63–78

Bodansky D (1993) The United Nations Framework Convention on Climate Change: A commentary. *Yale Journal of International Law* 18(2):451–558

Bohringer C and Vogt C (2004) The dismantling of a breakthrough: The Kyoto Protocol as symbolic policy. *European Journal of Political Economy* 20(3):597–617

Boyd E G K (2003) "Forests post-Kyoto: Global priorities and local realities." Unpublished PhD thesis, University of East Anglia

Carrere R (2000) Sinks that stink. *World Rainforest Movement Bulletin* June

Foster J B and Magdoff F (2009) *The Great Financial Crisis: Causes and Consequences.* New York: Monthly Review Press

Fry I (2002) Twists and turns in the jungle: Exploring the evolution of land use, land-use change and forestry within the Kyoto Protocol. *Review of European Community & International Environmental Law (RECIEL)* 11:2

Fry I (2007) More twists, turns and stumbles in the jungle: A further exploration of land use, land use change and forestry decisions within the Kyoto Protocol. *Review of European Community & International Environmental Law (RECIEL)* 13:1

Geertz C (1963) Agricultural Involution: The Process of Ecological Change in Indonesia. Association of Asian Studies, Berkeley: University of California Press

Goldenweiser A (1936) Loose ends of a theory on the individual pattern and involution in primitive society. In R Lowie (ed) *Essays in Anthropology Presented to A. L. Kroeber* (pp 99–104). Berkeley: University of California Press

Grubb M with Vrolijk C and Brack D (1999) *The Kyoto Protocol: A Guide and Assessment.* London: The Royal Institute of International Affairs

Grubb M and Yamin F (2001) What happened at The Hague, why, and where do we go from here? *International Affairs* 77(2):261–276

Gutiérrez M (2007) "All that is air turns solid: The creation of a market for sinks under the Kyoto Protocol on Climate Change." Unpublished PhD thesis, City University of New York

Hajer M A (1995) *The Politics of Environmental Discourse: Ecological Modernization and the Policy Process.* Oxford: Oxford University Press

Harvey D (1996) *Justice, Nature and the Geography of Difference.* Oxford: Blackwell

Harvey D (1999 [1982]) *The Limits to Capital.* Oxford: Blackwell

Harvey D (2001) *Spaces of Capital: Towards a Critical Geography.* New York: Routledge

Harvey D (2003) *The New Imperialism.* Oxford: Oxford University Press

Harvey D (2006) *Spaces of Global Capitalism: Towards a Theory of Uneven Geographical Development.* London: Verso

Henwood D (1997) *Wall Street: How it Works and For Whom.* London: Verso

Intergovernmental Panel on Climate Change (IPCC) (2000) *Land Use, Land Use Change and Forestry.* Cambridge: Cambridge University Press

IPCC (2003) *Good Practice Guidance for Land Use, Land Use Change and Forestry.* Cambridge: Cambridge University Press

Jackson R B, Banner J L, Jobbágy E G, Pockman W T and Wall D H (2002) Ecosystem carbon loss with woody plant invasions of grasslands *Nature* 418 (8):623

Katz C (1998) Whose nature, whose culture? Private productions of space and the 'preservation' of nature. In N Castree and B Braun (eds) *Remaking Reality: Nature at the Millennium* (pp). New York: Routledge

Latour B (1999) *Pandora's Box: Essays on the Reality of Science Studies.* Cambridge: Harvard University Press

Lund H G (1999) *Definitions of Forest, Deforestation, Afforestation and Reforestation.* Manassas: Forest Information Services

Newell P and Paterson M (2010) *Climate Capitalism: Global Warming and the Transformation of the Global Economy.* Cambridge: Cambridge University Press

Ott H (2001) Climate change: An important foreign policy issue. *International Affairs* 77(2):277–296

Ribot J C (1998) Theorizing access: Forest profits along Senegal's charcoal commodity chain. *Development and Change* 29:307–341

Robertson M M (2006) The nature that capital can see: Science, state, and market in the commodification of ecosystem services. *Environment and Planning D: Society and Space* 24:367–387

Schapiro M (2010) Conning the climate: Inside the carbon-trading shell game. *Harper's* February

Schlamadinger B and Marland G (2000) *Land Use and Global Climate Change: Forests, Land Management and the Kyoto Protocol*. Washington, DC: Pew Center on Global Climate Change

Smith N (2008 [1984]) *Uneven Development: Nature, Capital and the Production of Space*. Oxford: Blackwell

Yamin F and Depledge J (2004) *The International Climate Change Regime*. Cambridge: Cambridge University Press

Yengoyan A A (2001) Culture and power in the writings of Eric R. Wolf. In E R Wolf *Pathways of Power: Building an Anthropology of the Modern World* (pp). Berkeley: University of California Press

Chapter 4
Between Desire and Routine: Assembling Environment and Finance in Carbon Markets

Philippe Descheneau and Matthew Paterson

Introduction

> Why choose Bluenext? We are the fastest exchange. It takes only 15 minutes to settle a trade...Being fast is important as the environment needs a speedy solution (advertisement for Bluenext, an exchange platform for carbon trading).

Carbon markets have grown rapidly and have become a core element in the policy response to climate change. But how they have been constructed is not yet well understood. Promoters of carbon markets tend to assume a "natural" roll-out of a market logic, while critics quickly ascribe colonial or nefarious corporate intent (Bachram 2004; Lohmann 2005; Smith 2007). Our aim in this article is to go beyond such simplistic accounts and to contribute to an emerging literature that takes seriously the constructed character of markets, including carbon markets.

Historically, carbon markets, like other environmental markets, were conceptualised as means to internalise the costs created by pollution.[1] These economic models may, as MacKenzie claims in the case of finance (2006), be performative, but this fact says little about how the details of market construction play out. In fact the models only tend to describe and analyse the policy tools, not the market that results. This rests on an implicit assumption that the two are the same.

The New Carbon Economy, First Edition. Edited by Peter Newell, Max Boykoff and Emily Boyd.
© 2012 Philippe Descheneau and Matthew Paterson. Book compilation © 2012 Editorial Board of Antipode and Blackwell Publishing Ltd.

Most of the work on carbon markets has thus focused on the role of the state (or EU institutions in the case of the European Union Emissions Trading Scheme, or EU ETS) in creating those markets through regulation or the explicitly political dynamics surrounding them (Aldy and Stavins 2008; Skjaerseth and Wettestad 2008; Victor and House 2004; Voss 2007). Even critics of carbon markets tend to focus on the narrowly political questions surrounding carbon markets, such as the way that carbon offsets are governed (Bumpus and Liverman 2008) or the legitimacy crises they produce (Paterson 2010). This focus tends to reify the market as a natural social institution and that the only work of social construction involved is in the policy-making process. For example, in the Clean Development Mechanism (CDM), the basis of the markets we focus on here, there is a substantial literature detailing the complex bureaucratic process involved in moving from an initial project idea through to the issuance of a Certified Emission Reduction unit (CER), the commodity created in the CDM process (Michaelowa and Jotzo 2005; Paulsson 2009; Streck 2004; Yamin and Depledge 2004). This process entails a number of stages and actors, including market actors, but the literature focuses mostly on the activities of the UN bodies involved, specifically the CDM's Executive Board and Methodologies Panel.

The Cultural Construction of Carbon Markets

We contend here that there is in fact an enormous gap between a policy mechanism like the CDM and the markets which are constructed around it. There are many processes involved in the passage from one to the other. There is an embryonic literature focusing on this type of question (Callon 2009; MacKenzie 2009a; Knox-Hayes 2008a, 2008b; Lövbrand and Stripple 2008; Paterson and Stripple 2010), but there is still much work that could be done, and this article is one attempt to add to our understanding of these processes.

The approach we adopt here, broadly consistent with this emerging literature, is through theoretical tools that are offered by the field of cultural political economy. By this we follow Best and Paterson's account (2010) rather than that of Jessop and Sum (2001; Jessop 2004; Jessop and Oosterlynck 2008; Sum 2005). For these latter authors, cultural political economy is regarded as a specific approach, a "cultural approach to political economy" (Jessop and Sum 2001:98), which in effect (despite protestations of pluralism) insists on Marxism as the basis for the political economy and then interrogates how culture, understood as semiotic processes (Jessop 2004), constitutes specific patterns of practice in capitalist economies.

By contrast, for Best and Paterson (2010), the aim is as much to politicise the literature on cultural economy (Amin and Thrift 2004; du Gay and Pryke 2002; Ray and Sayer 1999); that is, to insist on the importance of questions of power and authority in mediating how culture constitutes economic processes. However, this intent is pluralistic in how theoretically such a project may be pursued, and it thus shares with the cultural economy literature a theoretical pluralism, understanding it not as a specific approach but rather a field with various approaches (including those that Jessop and Oosterlynck 2008 dismissively term "soft economic sociology") sharing an insistence on the importance of culture as a concept to understand the construction of the economy. In broad theoretical terms, it can be understood as operating in the terrain, or debate, between poststructuralism, especially in this context actor-network theory (ANT) (eg Callon 1998), and more classical but culturally inflected accounts of political economy drawing in particular on Marx and Polanyi (see Best and Paterson 2010 for a further elaboration). Broadly this field can be understood as a debate between those emphasising the broad, historically constituted structures that serve to shape subjectivities and practices at specific moments in time and space, and those more wary of making such broad claims about historical structures. But these approaches share an insistence on showing how culture—understood specifically as the intersubjectively produced meanings through which practices are rendered intelligible and normatively charged—is crucial to understanding political economy (Best and Paterson 2010:12–17).

In the context of debates about carbon markets, this field can be translated into those between ANT scholars and eco-Marxists. Most of the literature on carbon markets within what we term cultural political economy to date have been carried out within ANT frameworks (Callon 2009; Knox-Hayes 2008a, 2008b; MacKenzie 2009a) or drawing on Foucauldian notions of governmentality (Lövbrand and Stripple 2006; Paterson and Stripple 2010); the exceptions are Bumpus and Liverman (2008) and Paterson (2010). More broadly in relation to environmental commodification, however, eco-Marxist analyses are widespread (see in particular Castree 2003, 2005; Heynen et al 2007; Mansfield 2008). While there is significant critical debate between these two approaches,[2] our interest is not to take a position in this debate, but rather to take as a point of departure how both enable empirical research focused on the specific cultural processes through which carbon markets are constructed. In ANT studies of economic processes, this is the overall purpose (eg Callon 1998; MacKenzie 2006, 2009b), while in eco-Marxist analyses, there may be the goal of identifying the various means by which neoliberal, or finance-led capitalism, is being (re)produced and contested in general (Jessop 2004) or in the context of environmental markets specifically (see especially Robertson 2007). But both types of approach emphasise the role of subject formation (Paterson and Stripple 2010; Thrift 2001), the performativity of economic discourse (Callon 1998; MacKenzie 2006, 2009b; Robertson 2007), the emergence of specific types of practice, notably forms of calculation (Miller 2001), of technologies and non-humans as *actants* (Knorr-Cetina and Bruegger 2003) and the interpretive practices of economic actors (Ouellet 2010; Walters 2010).

For ANT researchers, this might be the end in itself, while for eco-Marxists, this serves to demonstrate how structural economy shapes these practices and economic discourses. In this article we want to show how what are assumed to be more technical processes can have important political effects and how culture can help to shape it. We highlight two elements of such processes in carbon markets, and the way they interact in the construction of those markets. The first of these is the generation of affective processes by which market participants intersubjectively mobilise desire for carbon markets as a whole. The second is the sorts of borrowing and translating financial practices from other such markets that market participants.

For the first of these, we draw in particular on Thrift's work on the construction of the "new economy" as a discourse.[3] Thrift (2001) argued that the elaboration of a set of practices around the new economy which tended to bring its discursive object into being involved in particular the creation of a "romance" of the new economy to which subjects could orient themselves. This romance was articulated through a number of means, termed by Thrift (2005) the "cultural circuit of capitalism", including technology and business gurus, MBA programmes, conferences and trade fairs, and popular magazines. This seems to us particularly fruitful in understanding the construction of carbon markets.

Drawing on Thrift's argument, we inquire into this social construction by exploring the construction of the subjectivities of carbon market actors. The mobilisation of the subjectivity of actors is key to creating the desire that animates market actors' practices, and is pursued through an intersubjective process of the creation of specific narratives around the carbon markets' purposes. We develop here one of the only studies of such subject formation in carbon markets (see also Paterson and Stripple 2010).

Second, we explore the way that carbon market actors borrow from existing financial practices to make the emerging market readily intelligible, to enable it to operate as a matter of financial routine. A key process in the construction of the carbon markets has been the emergence of a set of financial services (various derivative products, in particular) around the trading in the basic credits and allowances. For some, this process is simply a matter of the pursuit of efficiency in carbon markets, because such financialisation reduces transaction costs and enables actors to hedge against price volatility and other business risks (Knox-Hayes 2008b). However, this process has to be understood, historically and politically, in the context of the (re)financialisation of the economy that has taken place since the late 1970s (Aglietta 1999; Helleiner 1994; Leyshon and Thrift 1997), which has translated into the emergence of carbon markets in numerous ways. The currently dominant financial market actors have mobilised their resources and practices in the construction of carbon markets; it is this that needs the analysis to be a cultural *political* economy rather than simply a cultural economy approach.

This part of the article takes some insights from ANT in the way an actor-network is constituted. In particular, we focus on some of the technical devices through which carbon market practices become routinised, which in ANT terms could be treated as "actants", non-human elements in a network that nevertheless exert "actor-like" effects. Social studies of finance (Callon 1998; Callon, Millo and Muniesa 2007; Knorr-Cetina and Bruegger 2003, 2004; Mackenzie 2006, 2009a, 2009b; Preda 2003) focus on these types of devices. Mackenzie and Hardie (2009) analyse for example the assembling of an economic actor by looking at hedge fund practices. Knorr-Cetina and Bruegger's account (2003, 2004) of the role of devices in foreign exchange markets also tries to understand the role of socio-technical devices in markets. Technology has in this case an important role in changing, from networks to the screen, the way prices are discovered or established. Preda (2003) also discusses the role of technology, from the ticker to the screen in the historic development of stock exchange. Borrowing practices from existing financial markets, as we suggest in more detail below, can be seen to have operated in such a way.

In elaborating these cultural processes in the construction of carbon markets, we focus on the markets that have emerged around the CDM. The CDM markets illustrate these processes and dynamics clearly. Our focus is thus more on the importance of the apparatus than technology per se as the market is still in the early stage of the in vivo experiment (Callon, Millo and Muniesa 2007). On the one hand it entails all sorts of highly routinised and bureaucratised processes, from the approval processes in the UN machinery to the decision processes by investors about profitability and various sorts of risks. On the other hand the CDM involves all sorts of actors (including those in the UN process) presenting the CDM as a market opportunity, and participation in CDM projects has to be discursively constructed as a desirable thing to do through discourse.

Mobilising Desire

In this section, we focus on how carbon markets are mobilised through a sort of affective desire, through what might, following Thrift (2005), be called the "cultural circuit of carbon markets". These have been central to the generation of an orientation towards such markets which is not simply motivated by calculations of profit and risk, but is mobilised by a sort of liminal energy channelling through the boosters of these markets. It is definitely, as Thrift would elsewhere put it, the "romance, not the finance" which makes carbon markets go round.

Carbon Expo is the "Global Carbon Market Fair & Conference", where around 3000 carbon market participants meet annually.[4] What is striking when you attend Carbon Expo is the energy and enthusiasm for carbon markets which animate their proponents and participants, and at the same time the occasional anxiety (an anxiety more marked in 2010 than in previous years) that the whole rapid development of the market may be sustained largely by that energy and enthusiasm.[5] It operates in part through a collective re-affirmation of the simultaneous environmental beneficence and commercial promise of carbon markets, and a validation of the inventiveness of the market actors in precisely their skill in *bricolage*; their ability to put different actors, techniques and products together in myriad ways to manufacture value. It is at the same time an immense networking site, where actors involved in carbon markets do deals and construct relationships on which later deals may be based. Presentations frequently have the feel of a marketing spiel, except that the marketing is for the market as a whole as much as for any specific product or firm.

Carbon Expo is also an opportunity for actors to profess their faith in the market and the financial sector.[6] The development of market products (registries, software, risk hedging, financial products, standards, rating agencies, specific abatement products or technologies) might look somewhat overdeveloped given some of the uncertainties surrounding the future markets, but the majority of the participants are convinced that the market can only grow (IETA 2008). Carbon Expo also allows actors to get a sentiment of the market[7] and a glimpse at future directions in regulation and finance (such as likely developments of the CDM processes, the likely timescale for a carbon market in North America, or new possible markets such as in forestry and agriculture) and is presented as a key site for the dissemination of financial innovations.

Carbon Expo may look like other market fairs given the hype and energy for new products. But there is a major difference in the products being sold. While new products such as the iPad that are clearly hyped enormously, the hype has some relationship to the (purported) use value of the object. By contrast, the products in the carbon market have no use value. The tonne of carbon refers to a tangible unit of measure, but demand for the rights to emit it arise purely out of government regulatory activity. The tonne of carbon has thus to be abstracted to something more tangible for market actors, ie financial or monetary products. Thus, what is being sold is not the tonne per se but rather the financial or discursive representations of it.

Thrift (2010) argues that we are currently in a period where capitalism is generating value and growth through tapping particular cultural veins:

> That vein is what I will call the technology of cultural composition, in that it involves charging up the semiotic sphere in order to create and tap bodies of passion which display talent . . . The result is clear at least: a vast man-made imagination machine, but a machine bent on directing imaginations in particular ways by multiplying promise and boosting potentiality (Thrift 2010:198).

This seems to us an apt interpretation of driving forces in carbon market development. Carbon markets actors display precisely this passion on which their work depends. Carbon markets depend not only on a set of normative value-orientations towards climate change but more generally on an enthusiasm precisely for the novelty of such markets. Innovation is an end in itself for many in the carbon markets: a new type of project, a new methodology, a new means of selling CERs, a new market coming on stream, constant reflection on the design of markets. In fact, for many, this is far more important as a driving orientation than a set of values around climate change.

Nevertheless, tapping passions for climate change is central in constructing demand in these markets. Marketing offsets invokes and helps to construct a series of consumer subjectivities, at the most general level a guilt around their emissions levels, but transformed into a motivation to "do good" by investing in offset projects, especially in the South, sometimes with specific combined ends such as deforestation.[8] And the initial guilt motivation can be channelled in particular directions, as in the notion of "going on a low carbon diet", which enables carbon marketers to tap into specific sorts of orientations towards consumption, those motivated by the extreme precision of calculation (carbon and carbs), the relationship between desire, denial, and the "treat", and so on (see Harrington 2008; Paterson and Stripple 2010).

For the bulk of demand (particularly in the voluntary markets but also in the EU ETS-led CDM market where firms meet regulatory obligations under the EU ETS but also use CDM purchases for marketing purposes), there is a more complex semiotic chain. Offset firms tap into the desire amongst purchasing firms (the vast majority of demand in the voluntary markets is corporate not individual) not only at the level of individual managers' own subjectivities, but in their desire for "green Public Relations". The pursuit of carbon offsets has become a strategy to motivate both employees and consumers, through notions such as carbon neutrality. This is a means by which they can then transfer the passion of their consumers into a marketing opportunity, mediated or facilitated by the offset markets. For financial firms like HSBC, it then can also stimulate an interest in becoming players in the carbon markets themselves. HSBC became the first major bank to claim to be carbon neutral in October 2005 and decided to do much of its carbon neutrality projects "in house" (Bayon, Hawn and Hamilton 2007:153) in order to build its carbon trading and investment capacity.

Examining advertising within the carbon markets is useful in this context.[9] This is not to see if it involves greenwashing (Beder 1997; Greer and Bruno 1996), but rather to observe how the carbon economy is constituted discursively. What is striking is that much of this advertising focuses on the normativity of climate change rather than the opportunity of profits. This is surprising given the audience for this advertising is not the general public. They are rather aimed at other firms in the carbon markets, verifiers advertising to project developers, lawyers advertising to carbon finance firms, and so on.[10] This is not to suggest that the readers of these ads may be less swayed than a more general audience by clever marketing; our point here is that this nevertheless is more useful in telling us something about the shared discursive universe of carbon market actors than would be the case for advertising to general consumers. This thus represents one way of looking at how carbon market actors talk to each other.[11]

Much of this publicity material contains simple banal images such as might be expected. First are images of "nature" animals, flowers, trees, mountains, glaciers and icebergs. Some of these are climate specific, especially the frequent deployment of ice looking like it might be melting, while others are more generic.

Second, there are images of various technologies associated with "clean development" especially wind turbines and solar cells, but also again trees (these get framed as both nature and technology, depending on context). At times there are images of more conventional factories with smokestacks, representing "dirty" development which needs carbon finance to help cleanse it.

These banal images do some simple work. They establish the basic environmental legitimacy of carbon markets and their various practices. They help to produce or mobilise one element in the desire of animating carbon markets, that of the normativity of "saving the planet". Again, it is interesting to note the ubiquity of these sorts of images in the context of ads aimed exclusively at the actors involved in the carbon markets themselves, not at "consumers" in the conventional sense.

Figure 1: BlueNext "be in safe hands" (source: reproduced with kind permission of Bluenext)

But some of these images of "nature" go beyond the banal, and start to do more complex semiotic work. These tend to be focused around an image of the planet as a whole. In particular, they frequently conjure up images of security of the planet as something to be secured. This is most obvious in this image from BlueNext, the exchange established by NYSE Euronext and Caisse des Dépôts (Figure 1). This ad deploys an image of the earth, a classic in environmental discourse since the late 1960s,[12] cradled in the "safe hands" of the white businessman. The notion that we now live in the anthropocene (Dalby 2007) is ably illustrated here while the man's hands work to make the earth safe; it is nevertheless clear that it is he who is now dominant in shaping the earth's future.

Other ads pursue similar images. An ad by APX, "leading infrastructure provider" for carbon markets, that is, they work on the registries, inventories, and management systems, similarly shows the earth as an entire system, but the security metaphor is more subtle. In that ad, the earth is figured in a drop of water on a leaf, conveying the image that the earth is there in the smallest object, by extension is reflected in every activity, and requires intimate management to be secured. The slogan under APX's logo is "Energy, Environment, Market Integrity", and the text in the ad focuses on "management" throughout, important in sustaining this essentially biopolitical signification (Lövbrand and Stripple 2006).

Security is pursued in carbon market marketing in other simpler contexts. Most obvious is in the name of some firms, EcoSecurities being the obvious one, which gives us the double meaning of ecological security and securities as financial instruments. But security is also closely linked to reliability or trustworthiness in adverts in the carbon market. This is clear in the case of the APX ad, where integrity "keeping things whole" is the means of indicating that the company is serious in pursuing climate security.

Security and reliability operate fairly obviously as themes in advertising in the carbon market. They enable the emphasis on pursuing environmental goals. But they also enable carbon market actors to present themselves to each other in terms of safety and reliability. In a context where there is legitimacy of both financial actors in general, given the crisis which started in 2007, with its origins in various "creative" financial practices,[13] and of carbon markets, with various scandals about carbon offset projects, and more recently tax fraud by carbon traders (Paterson 2010; Reuters 2009), the emphasis on such themes is striking, and we would interpret it in part as efforts to shore up the legitimacy of carbon markets.

But security and reliability are not enough; carbon market actors must be motivated. In these ads at least, however, this is not principally done through the desire for performance. As suggested above, the most common way is through the standard normative arguments about climate change. However, the way that performance operates is through a link to this, specifically through the theme of speed. As BlueNext puts it in their brochure, "being fast is important as the environment needs a speedy solution". An ad by SGS which they use widely in carbon markets exemplifies this theme. The ad figures a snow leopard on a white background moving determinedly towards the reader, with the slogan "Speed up Emissions Verification, Please!" underneath. The snow leopard is taken to be exhorting us to speed up the verification process. The leopard is perhaps also deployed to represent the threat of climate change itself; it may say please but we will experience nature as "red in tooth and claw" if we fail on climate change. On the back of the flyer, SGS states "faster, more accurate emissions verification is better for business and better for the planet".[14]

All these ads can be understood as general attempts to motivate subjects, both with regard to the services of particular firms, and also in relation to carbon markets themselves. One last ad, for Climate Change Capital, a carbon finance firm established by climate change lawyer and former NGO-activist-cum-diplomat for small island states James Cameron, makes this explicit. Its slogan is "creating wealth worth having", and like many such slogans, is best understood through its negation. All this other wealth is not worth having, it implies. In other words, wealth (as money, capital, etc) is to be fundamentally distinguished from value, or worth. The point of carbon finance is to close that gap to create wealth that actually has value.

Borrowing Financial Practices

So carbon markets are mobilised in part through the intersubjective production of desire. But desire is not enough. Alongside desire there must be technical knowledge, and more importantly routinised practice. This is in large part because the professional training and discipline of financiers necessarily leads them to understand their practice in general in terms of the tenets of orthodox finance theory. But it has the effect of channelling desire in highly efficient ways. This paradoxical relationship between desire and economic rationality has been extremely effective in mobilising carbon markets.

The ads do some of this work. They are, by comparison with most advertising, very text heavy, conveying technical information about the financial or other services being offered. But much of this routinisation, or even banalisation, of carbon markets, has occurred through borrowings from other areas of financial activity. We present two examples of this borrowing here, although many others could be identified.[15]

The first is the development of a credit rating agency in carbon markets. Credit rating agencies have become key players in many financial markets, providing quick indications of the agency's judgment of the credit-worthiness of firms, governments, or other institutions, and simplifying investment and lending decisions for financiers (see Sinclair 2005). They also produce a form of simplified market structure, with access to capital determined to a considerable extent by the rating given by Standard & Poor's or Moody's. Within carbon markets, IDEAcarbon is trying to mimic this by developing a credit rating agency for the carbon markets, now called simply the Carbon Rating Agency.[16] The rating system is borrowed directly from those existing in other financial markets, with ratings from AAA down to D. The ratings are intended to provide investors with a set of third party expert judgements about the riskiness of different types of CDM or joint implementation (JI) projects. As a consequence, business models underpinning projects as well as overall firm strategies would start to coalesce around a limited number of types produced by varying orientations to three different types of risk (validation, registration and volume) to do with getting projects through the CDM cycle or selling on CERs in the secondary market. This decision process centres around the relationship between risk and reward—investing earlier in a project cycle means greater potential reward, as the investment occurs when prices are low, realising gains in the secondary CER market where prices are higher, but entails higher risk, to do with project failure but also regulatory risk (the CDM approval process). The project is only young (starting in early 2008) and cannot be said to have achieved these results but the intent is clearly to shape *what the market looks like*.[17] The intent is to create a situation where CER prices coalesce around specific poles, reflecting different types of risk/reward relationship. As the Agency puts it, the aim of credit rating is to "help establish a clearer relationship between the price of carbon and the delivery risk, *thus helping the carbon market to mature*" (emphasis added). This is manifestly a construction of the market rather than a simple reflection of a "naturally existing" market logic, as it attempts to shift from a situation where the price spread for CERs is wide and complex, making it difficult for investors to discern patterns in why some types of investments command higher prices than others, to one with four easily identifiable types of business model with specific risk–reward profiles. In other words, it involves constructing a market structure which facilitates simple, routinised decisions by investors, rather than ones where the investor has to start from scratch each time.

The second borrowing concerns the range of financial derivative instruments in carbon markets, borrowed directly from such instruments in other financial markets. A booklet produced by Barclays Capital, "Emissions Trading at Barclays Capital" (Barclays Capital 2008), illustrates this effectively. In that document, alongside various sorts of market analysis and description of Barclays' services, is a detailed account of the sorts of "risk management structures" available to clients operating in carbon markets, in particular in those markets that are "fully commodified".[18] This includes a description of the full range of derivative products (options, futures, swaps), but then further detail on several of them. For example, there is a discussion about the means of transforming CERs into EUAs,[19] both the means Barclays Capital propose (a straightforward swap with a €7.00 cash payment reflecting, although of course cementing, the rough price differential between the two), as well as the strategies to adopt given the price difference between CERs and EUAs—the flat "forward curve" for CER prices, and that these allowances can be banked between different years in the scheme's operation. Thus the "2008 EUA 2012 CER optimal swap" is to buy 2012 CERs in 2008, and wait until 2012 to transform them into EU emission allowances (EUAs). Barclays Capital also outlines what they call a "zero-cost collar structure". This is designed to enable clients to hedge against price movements (up or down) without paying a premium, and involves a structured relationship between a call option by the client and a put option by Barclays Capital (see Figure 2). The document outlines a number of other such arrangements. Our point is not to get

Zero-Cost Collar Structure

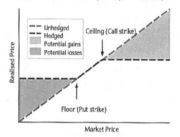

Description

- A collar is a zero premium structure that allows a client to hedge against upside and downside price movements. The client buys an EUA call option with a strike price higher then the strike price of a put option it sells.

Key Terms	
Commodity	EUA/CER
Call Option Buyer	Client
Put Option Buyer	Barclays Bank PLC
Call Option Strike	€/tonne
Put Option Strike	€/tonne
Expiry Date	5 business days before delivery
Delivery Date	1ˢᵗ Dec
Settlement Date	20th calender day of the month following delivery.

Benefits

- A collar caps the purchase price of the EUA. The client pays no premium for this structure as the purchase of the call is funded by the sale of the put.

- If the market price falls below the floor and the put is exercised against the client, the EUA price will still be cheaper then the current market price.

Risks

- Market price is below strike, and option expires worthless (out of the money)

BARCLAYS CAPITAL

Figure 2: Barclays Capital's "zero-cost collar structure" (source: reproduced with kind permission of Barclays Capital)

too much into the details of any particular scheme, but to draw out a more general point. These models are on the one hand direct borrowings from existing financial markets, adapted to particularities of emissions trading systems (in particular, the character of permits as property rights, the requirement to surrender permits at specific dates, and the linking mechanism between the EU ETS and the CDM), and on the other hand, actively construct the market that emissions trading proponents suggest exists "naturally". In particular, as with the ratings agency, they render investments in carbon markets a matter of routine for financiers, rather than ones that they need to generate distinctive expertise in.

Conclusion

Together, these two aspects of the construction of carbon markets, desire and routine, can be captured well by two points made very effectively by Nigel Thrift in relation to the "new economy". First, Thrift suggests that "Effective social movements need to create background, a taken-for-granted world which, if you like, assumes the new economy's assumptions" (Thrift 2005:117). This seems to us a useful analysis of the process occurring as carbon markets have developed. One of the effects of the establishment of this liminal energy around carbon markets, combined with their normalisation through the borrowing of financial practices, is that carbon markets have become common sense, the "taken-for-granted world". Even in the context of the uncertainties produced by the collapse of the UN negotiations in Copenhagen in December 2009, considerable problems in enacting legislation in the USA, and the impact of the economic crisis on the overall legitimacy of financial actors, what is striking is the continued enthusiasm of market actors for the market (Bernstein et al forthcoming). While at Carbon Expo 2010 there was much reflection on the decline in demand for primary CERs because of uncertainty about an international agreement to come into effect in 2013, on the ongoing problems of getting legislation through in the USA, and on the various scandals in the EU ETS in the previous couple of years, the overall "market sentiment" was still that the future of carbon markets is rosy.

Second is the basic point in the title of Thrift's article "it's the romance, not the finance..." In practice, in his article, the argument becomes more like "it's the romance *of* the finance". But the point is nevertheless apposite. From around 1997 onwards, carbon markets became romanticised by their protagonists. They became understood as a heroic, romantic effort to marry climate mitigation and economic growth. This romance has continued to date. Interviewing carbon market actors, or attending events such as Carbon Expo or IETA side events at a UNFCCC COP (United Nations Framework Convention on Climate Change Conference of Parties), one is struck by the affective economy of such markets, that they are driven and sustained as much by an emotional investment in carbon trading as in narrowly financial assessments of investment opportunities and strategies. Indeed, following Thrift's logic, it is this emotional investment, this romance, which creates the energies to engage in more "cold", "rational", investment strategies. In conversations with carbon traders, occasionally they get a moment of anomie where they worry that the whole thing is sustained solely by their enthusiasm, which is simultaneously for the money making opportunities and for the contribution to greenhouse mitigation (and frequently, for the development benefits of things like CDM or voluntary market projects).

We have tried here to contribute to the emerging studies focusing on the social and cultural construction of carbon markets. We have not tried to resolve the many debates that divide different approaches to this question, but rather to deploy specific concepts developed within this debate to understand the ongoing assembling of carbon markets. We have suggested that one particular way to understand this process is through the relation between the construction of desire for carbon markets and the borrowing of financial practices that enable financiers to routinise their carbon investment or trading practices. Clearly, theoretical divides could be highlighted. In particular, eco-Marxists could clearly start by emphasising that the social and cultural construction of carbon markets has to be understood in the context of the financial power in the economy. In contrast, ANT scholars might focus on how the process of constructing carbon markets shows how that process can be fragile, subject to contestation, and rarely represents a neat "roll-out" of financial power.

There are clearly some limits to the claims we can make here. There is clearly space for more exhaustive analysis both of advertising and of the borrowing of financial practices in carbon markets. Nevertheless, we do want to claim that the evidence we have presented shows that an important element in the construction of carbon markets is through a relationship between desire and routinisation. The desire to do good (both environmentally and financially) and the routinisation of financial practices applied to the carbon markets are thus vital elements of the market. But this relationship is also to some degree in tension. The legitimising dimension of environment in the context of a profit-making business is constantly stressed at the same time that carbon is being treated as a banal commodity. As a consequence, the door is opened to the critiques of carbon markets that suggest they are simply a form of greenwash, a cooptation of rhetoric about emissions reduction that in practice does nothing to achieve such cuts. But given the narratives that market actors tell about their own motivation (as evidenced in the advertising discussed above and at events like Carbon Expo), this tension plays out not only on the level of overt political conflict between proponents and opponents of carbon markets, but also within the identities of market actors themselves. The tension can thus be expected to play a role in how such actors construct markets on a daily basis. There are limits to the cognitive dissonance which would be produced by entirely ignoring the goal of emissions reductions while constantly telling a story about such reductions.

Acknowledgements

An earlier version of this paper formed part of a paper entitled "regulation and value in carbon markets", which was presented to a workshop on International Financial Standards and the Environment, at the University of Waterloo, 25 September 2009, and in a paper by Matthew Paterson given at the International Studies Association Annual Convention in New York, February 2009. We are grateful to the participants at both events for the lively discussion and enthusiastic feedback on the paper. The research has been financed under an award from the Social Sciences and Humanities Research Council of Canada, entitled "climate change politics and the greening of the state".

Endnotes

[1] Economic models for property rights on pollution advocated by different economics dates back to Coase (1960) and have been applied more recently in the case of carbon by Barrett et al (1992). For a useful history of these ideas, see Gorman and Solomon (2002).

[2] See for example the critiques of ANT by Fine (2005), Gareau (2005) or Castree (2002); the response by MacKenzie (2009b:32–33), and the attempt at a synthesis by Holifield (2009).

[3] We put new economy in quotation marks following Thrift's understanding of it as a discourse about the transformation of the economy through ICTs, rather than referring to any "real" changes in the economy itself.

[4] We have each been to Carbon Expo for fieldwork. Paterson attended the 27–29 May 2008 and again the 26–28 May 2010 conferences, both in Cologne; Descheneau attended the 27–29 May 2009 conference in Barcelona. Carbon Expo is the world's largest global carbon market fair and conference and is organised by the World Bank and the International Emissions Trading Association (IETA). In 2009, the seventh conference received 3000 visitors from 111 countries (CarbonExpo 2009). Methodologically, the research has involved extensive observation and informal conversations with participants.

[5] Indeed, the desire to be "cool" on the part of carbon market actors sometimes attains comic proportions. One workshop at the 2008 Expo was entitled "Pimp my ERPA", referring to the MTV show "Pimp my ride". In this case, the painfully self-aware title masked a dull talk about the sorts of problems and pitfalls involved in legal negotiations over the form of the Emissions Reductions Purchase Agreement (ERPA), the basic contract between a CDM project developer and an investor. Of course, this is also a site of tension between desire and routine, in that organisations like the International Emissions Trading Association (IETA) also attempt to standardise the form of ERPA contracts in order to facilitate efficient market development.

[6] "CARBON EXPO is clearly the premier market event. While other events are struggling to maintain attendance due to the financial crisis, CARBON EXPO is the "go to" event and has increased attendance. It's the place where market participants do business, an event that you can't afford to miss. CARBON EXPO is showing the market confidence by bringing people together to innovate for the post-2012 market" (Anthony Hobley, Partner, Norton Rose, in CarbonExpo 2009:6).

[7] "CARBON EXPO is the most important carbon event of the year. It is the best way to assess the sentiment of the carbon market, get updated with the numbers presented in the fabulous annual report and with this, prepare our company for the carbon market of the near future" (Marco Monroy, CEO and President, MGM International in CarbonExpo 2009:7).

[8] Forestry projects make up about 33% of the voluntary market as opposed to under 1% of the CDM market (Capoor and Ambrosi 2009:29).

[9] This analysis draws on material collected at Carbon Expo by the authors. It is supplemented by some material from company websites (see also Dalby and Paterson 2006). We chose a selection of 20 adverts from a much larger sample collected, from different types of carbon market actors, financiers, auditors, government agencies acting as designated national authorities (DNAs) within the CDM process to try to get a roughly representative sample of advertising within the carbon markets. We focused on those pieces of publicity which contained striking images or sharp slogans, since these are likely to have the most semiotic effect on audiences and thus contain the themes through which carbon market actors seek to motivate audiences and themselves. We don't claim that this provides the basis for an exhaustive analysis; that would be beyond the scope of this paper. But we do claim that the themes we identify are reasonably representative of those which appeared in the ads we have examined. We only produce a selection here visually, for reasons of space, although we describe more images than we can show.

[10] One relevant point here is perhaps to emphasise, although it is not the point of our overall article, that sociologically there are a substantial number of carbon market actors

who worked in environmental NGOs before moving into the financial sector. This may help to explain how in these ads, as in other sites like Carbon Expo, the desire to "do good" and the desire to make money are fused in complex ways.

[11] There are of course limits in using adverts to demonstrate processes of subject formation, and the evidence this gives us should be read in conjunction with the other evidence we provide here. Specifically regarding our claim that this is part of how carbon market actors talk to each other, and our surprise about the environmental content of the ads, we would need to know more about the production of the adverts to sustain the claim more forcefully. It also remains to be seen whether those ads would have been different in they were targeting a larger audience. It may be that these images were selected by ad agencies external to carbon markets with little input from the firms themselves, which is a qualification about the interpretation we make here. We are grateful to Virginia Haufler for pointing this limit out.

[12] The representation of the earth as a single object has been important in environmental advertising since the late 1960s. From that period, "for the first time our visual environment allowed us to imagine the planet as a single organism" (Wilson 1991:167), and facilitated its construction as fragile and to be protected.

[13] It is notable that of the ads surveyed, only Barclays Capital makes any reference to the creativity that finance has come to be associated with.

[14] A copy of this flyer can be found on SGS's website at: http://www.climatechange.sgs. com/sgs-climate-change-flyer-leopard-e-lores-en-09.pdf, accessed 24 September2010.

[15] Our choice of materials here was guided by a selection of materials gathered at Carbon Expo in our two visits there. While the exploratory character of the research places limits on how systematic a survey we can legitimately claim here, we chose one practice by consultants aiming to shape overall market structures (the rating agency), one practice by a trading firm (Barclays Capital), and one attempt to popularise or "banalise" carbon market discourse.

[16] On IDEACarbon, see generally their website at http://www.ideacarbon.com/ default.asp, viewed 4 February 2009. For the Carbon Rating Agency, see http://www. carbonratingsagency.com/home/index.html, viewed 4 February 2009.

[17] Informal conversations with IDEA Carbon representative, Carbon Expo, May 2008.

[18] We mean this here in the sense that financiers use the term traders at Barclays Capital or CantorCO2e who can quickly observe price movements for different carbon asset classes and engage in arbitrage on the basis of their expectations of those movements in the short term. This is in sharp contrast to the usage of the term commodification in the social sciences, where, we tend to refer to commodification as referring to any process of production for market exchange, mostly going back to either Marx or Polanyi. Financiers refer to a product as commodified only if it can be represented as something totally standardised and thus comparable to each other by quick visual signals (the classic TV screens we see used by city traders) displaying price movements between different products. In other words, only when they are fully commensurable are products commodities.

[19] Under the EU ETS, a company must have enough EUAs to cover its total emissions, for each year of the particular phase of the ETS. But it is allowed to transform CERs (or ERUs, the unit in the Joint Implementation mechanism in Kyoto, although the CDM has become much the bigger of the two institutions) into EUAs.

References

Aglietta M (1999) La globalisation financière [Financial globalisation]. In Centre d'études prospectives et d'informations internationales (CEPII) (ed) *L'économie mondiale 2000* (pp 52–67). Paris: La Découverte

Aldy J and Stavins R (eds) (2007) *Architectures for Agreement: Addressing Global Climate Change in the Post-Kyoto World*. Cambridge: Cambridge University Press

Amin A and Thrift N (eds) (2004) *The Blackwell Cultural Economy Reader*. Oxford: Blackwell

Bachram H (2004) Climate fraud and carbon colonialism: The new trade in greenhouse gases. *Capitalism, Nature, Socialism* 15:5–20

Barclays Capital (2008) *Emissions Trading at Barclays Capital*. London: Barclays Capital

Barrett S, Grubb M, Rolland K, Rose A, Sandor R and Tientenberg T (1992) *Study on a Global Scheme for Tradeable Carbon Emission Entitlements. Tradeable Entitlements for Carbon Emission Abatement* (Project INT/91/A29). Geneva: UNCTAD

Bayon R, Hawn A and Hamilton K (2007) *Voluntary Carbon Markets*. London: Earthscan

Beder S (1997) *Global Spin: The Corporate Assault on Environmentalism*. Devon: Green Books

Bernstein S, Betsill M, Hoffmann M and Paterson M (forthcoming) A tale of two Copenhagens: Carbon markets and climate governance. *Millennium: Journal of International Studies* 39

Best J and Paterson M (eds) (2010) *Cultural Political Economy*. London: Routledge

Bumpus A and Liverman D (2008) Accumulation by decarbonization and the governance of carbon offsets. *Economic Geography* 84:127–155

Callon M (ed) (1998) *The Laws of The Markets*. Oxford: Blackwell

Callon M (2009) Civilizing markets: Carbon trading between *in vitro* and *in vivo* experiments. *Accounting, Organizations and Society* 42(3–4):535–548

Callon M, Millo Y and Muniesa F (eds) (2007) *Market Devices*. Oxford: Blackwell

Capoor K and Ambrosi P (2009) *State and Trends of the Carbon Market 2008*. Washington DC: World Bank

CarbonExpo (2009) *CarbonExpo 2009 puts the carbon market front and centre on the road to Copenhagen*. Press release, 29 May. http://www.press1.de/wrapper. cgi/www.press1.de/files/kmeigen_kmpresse_1243690543.pdf (last accessed 2 April 2010)

Castree N (2002) False antitheses? Marxism, nature and actor-networks. *Antipode* 34(1):119–148

Castree N (2003) Commodifying what nature? *Progress in Human Geography* 27(3):273–297

Castree N (2005) *Nature*. London: Routledge

Coase R (1960) The problem of social cost. *Journal of Law and Economics* 3:1–44

Dalby S (2007) Anthropocene geopolitics: Globalisation, empire, environment and critique. *Geography Compass* 1(1):103–118

Dalby S and Paterson M (2006) Empire's ecological tyreprints. *Environmental Politics* 15(1):1–22

du Gay P and Pryke M (eds) (2002) *Cultural Economy*. London: Sage

Fine B (2005) From actor-network theory to political economy. *Capitalism, Nature, Socialism* 16(4):91–108

Gareau B J (2005) We have never been human: Agential nature, ANT, and Marxist political ecology. *Capitalism, Nature, Socialism* 16(4):127–140

Gorman H and Soloman B D (2002) The origins and practice of emissions trading. *The Journal of Policy History* 14(3):293–320

Greer J and Bruno K (1996) *Greenwash: The Reality Behind Corporate Environmentalism*. Penang, Malaysia: Third World Network

Harrington J (2008) *The Climate Diet: How You Can Cut Carbon, Cut Costs, and Save the Planet*. London: Earthscan

Helleiner E (1994) *States and the Reemergence of Global Finance: From Bretton Woods to the 1990s*. Ithaca: Cornell University Press

Heynen N, McCarthy J, Prudham S, and Robbins P (eds) (2007) *Neoliberal Environments: False Promises and Unnatural Consequences*. London: Routledge

Holifield R (2009) Actor-network theory as a critical approach to environmental justice: A case against synthesis with urban political ecology. *Antipode* 41(4):637–658

IETA (2008) *GHG Market Sentiment Survey*. Geneva: International Emissions Trading Association. http://www.ieta.org/ieta/www/pages/download.php?docID=2282 (last accessed 3 May 2010)

Jessop B (2004) Critical semiotic analysis and cultural political economy. *Critical Discourse Studies* 1(2):159–174

Jessop B and Oosterlynck S (2008) Cultural political economy: On making the cultural turn without falling into soft economic sociology. *Geoforum* 39(3):1155–1169

Jessop B and Sum N-L (2001) Pre-disciplinary and post-disciplinary perspectives. *New Political Economy* 6(1):89–101

Knorr-Cetina K and Bruegger U (2003) La technologie habitée. La forme de vie globale des marchés financiers [Inhabiting technology: The global lifeform of financial markets]. *Réseaux* 21(122):111–135

Knorr-Cetina K and Bruegger U (2004) Traders' engagement with markets: A postsocial relationship. In A Amin and N Thrift (eds) *The Blackwell Cultural Economy Reader* (pp 121–142). Oxford: Blackwell

Knox-Hayes J (2008a) *Constructing Carbon Market Spacetime: Implications for Neo-Modernity*. Working Papers in Employment, Work and Finance, School of Geography and the Environment, University of Oxford

Knox-Hayes J (2008b) *The Developing Carbon Financial Service Industry: Expertise, Adaptation and Complementarity in London and New York*. Working Papers in Employment, Work and Finance, School of Geography and the Environment, University of Oxford

Leyshon A and Thrift N (1997) *Money /Space*. London: Routledge.

Lohmann L (2005) Marketing and making carbon dumps: Commodification, calculation and counterfactuals in climate change mitigation. *Science as Culture* 14:203–235

Lövbrand E and Stripple J (2006) The climate as political space: On the territorialisation of the global carbon cycle. *Review of international studies* 32:217–235

Mackenzie D (2006) *An Engine, Not a Camera*. Cambridge, MA: MIT Press

Mackenzie D (2009a) Making things the same: Gases, emission rights and the politics of carbon markets. *Accounting, Organizations and Society* 34(3–4):440–455

Mackenzie D (2009b) *Material Markets: How Economic Agents are Constructed*. Oxford: Oxford University Press

Mackenzie D and Hardie I (2009) Assembling an economic actor. In D Mackenzie *Material Markets: How Economic Agents are Constructed* (pp 27–64) Oxford: Oxford University Press

Mansfield B (ed) (2008) *Privatization: Property and the Remaking of Nature–Society Relations*. Oxford: Blackwell

Michaelowa A and Jotzo F (2005) Transaction costs, institutional rigidities and the size of the clean development mechanism. *Energy Policy* 33:511–523

Miller P (2001) Governing by numbers: Why calculative practices matter. *Social Research* 68(2):379–396

Ouellet M (2010) Cybernetic capitalism and the global information society: From the global panopticon to a "brand" new world. In J Best and M Paterson (ed) *Cultural Political Economy* (pp 177–196). London: Routledge

Paterson M (2010) Legitimation and accumulation in climate change governance. *New Political Economy* 15(3)

Paterson M and Stripple J (2010) My space: Governing individuals through the carbon market. *Environment and Planning D: Society and Space* 28(2):341–362

Paulsson E (2009) A review of the CDM literature: From fine-tuning to critical scrutiny? *International Environmental Agreements: Politics, Law and Economics* 9(1):63–80

Preda A (2003) On ticks and tapes: Financial knowledge, communicative practices and the rise of financial information technologies. Paper presented at the annual meeting of the American Sociological Association, Atlanta, GA, 16 August. http://www.allacademic.com/meta/p107749_index.html (last accessed 6 April 2011)

Ray L and Sayer A (eds) (1999) *Culture and Economy: After the Cultural Turn*. London: Sage

Reuters (2009) JPMorgan to buy EcoSecurities for $204 million. 14 September. http://www.reuters.com/article/GCA-GreenBusiness/idUSTRE58D37020090914 (last accessed 13 October 2009)

Robertson M M (2007) Discovering price in all the wrong places: Commodity definition and price under neoliberal environmental policy. *Antipode* 39(3):500–526

Schneider L (2007) *Is the CDM Fulfilling its Environmental and Sustainable Development Objectives? An Evaluation of the CDM and Options for Improvement*. Berlin: Öko Institute

Sinclair T J (2005) *The New Masters of Capital: American Bond Rating Agencies and the Politics of Creditworthiness*. Ithaca, NY: Cornell University Press

Skjærseth J B and Wettestad J (2008) *EU Emissions Trading: Initiation, Decision-making and Implementation*. Aldershot: Ashgate

Smith K (2007) *The Carbon Neutral Myth. Offset Indulgences for your Climate Sins*. Amsterdam: Transnational Institute/Carbon Trade Watch

Streck C (2004) New partnerships in global environmental policy: The Clean Development Mechanism. *Journal of Environment and Development* 13:295–322

Sum N-L (2005) *The Cultural Turn in Economics: Towards a Cultural Political Economy*. Cheltenham: Edward Elgar

The Carbon Rating Agency (2008) *The Carbon Rating Agency Report*. London: The Carbon Rating Agency

Thrift N (2001) "It's the romance, not the finance, that makes the business worth pursuing": Disclosing a new market culture. *Economy and Society* 30(4):412–432.

Thrift N (2005) *Knowing Capitalism*. London: Sage

Thrift N (2010) A perfect innovation engine: The rise of the talent world. In J Best and M Paterson (eds) *Cultural Political Economy* (pp 197–222). London: Routledge

Victor D F and House J C (2004) A new currency: Climate change and carbon credits. *Harvard International review* summer:56–59

Voss J-P (2007) Innovation processes in governance: The development of "emissions trading" as a new policy instrument. *Science and Public Policy* 34(5):329–343

Walters W (2010) Anti-political economy: Cartographies of "illegal immigration" and the displacement of the economy. In J Best and M Paterson (ed) *Cultural Political Economy* (pp 113–138). London: Routledge

Wilson A (1991) *The Culture of Nature: North American Landscape from Disney to the Exxon Valdez*. Between the lines: Toronto

Yamin F and Depledge J (2004) *The International Climate Change Regime: A Guide to Rules, Institutions and Procedures*. Cambridge: Cambridge University Press

Chapter 5
Ecological Modernisation and the Governance of Carbon: A Critical Analysis

Ian Bailey, Andy Gouldson and Peter Newell

Introduction

Ecological modernisation (EM) theories are centrally concerned with the relationship between environment and economy and with social capacities to recognise and respond to existing and emergent environmental problems (see Gouldson, Hills and Welford 2008; Mol, Sonnenfeld and Spaargaren 2009). EM theories therefore seek to describe, and in some cases to promote, the processes of framing and learning that might allow modern societies to mediate the relationship between environment and economy more effectively.

EM theories have helped to describe the ways in which environmental problems come to be framed as issues that are politically, economically and technologically solvable within the context of existing institutions and power structures and continued economic growth (Murphy and Gouldson 2000). This faith in science and technology, in governments and policy and in markets and economic growth creates a political space within which institutional learning and policy reform can more readily take place because it avoids triggering opposition from mainstream actors and established vested interests (Giddens 2009).

In this chapter, we argue that there are clear parallels between EM theories and contemporary mainstream debates on climate change. Consequently, an exploration of critical perspectives on EM theories can help us both to understand the processes through which the seemingly intractable problem of climate change has, over a relatively short period

The New Carbon Economy, First Edition. Edited by Peter Newell, Max Boykoff and Emily Boyd.
© 2012 Ian Bailey, Andy Gouldson and Peter Newell. Book compilation © 2012 Editorial Board of Antipode and Blackwell Publishing Ltd.

of time, been reframed as an opportunity to construct a new carbon economy and to anticipate some of the tensions, contradictions and limits of such an approach.

The chapter is structured in the following way. In the next section, we examine key features of EM theories. We argue that although the processes of framing and learning that are at the heart of EM make "wicked" (Rittel and Webber 1973) environmental problems such as climate change more tractable, they largely overlook the extent to which climate change can be dealt with without a change in social values or in society–nature relations. We also argue that they underestimate the procedural and distributional justice implications of this framing of the problem and the structural causes and consequences of a growth logic that may ultimately be fundamentally incompatible with dealing with climate change.

We then apply insights from EM theories to two cases that exemplify the emergence of market-based approaches to climate change, emissions trading and the market in carbon offsets. In each case, we explore the governance dimensions of these novel market mechanisms. We explore, the *political drivers* of these new forms of carbon governance: how neoliberal approaches to climate governance came to be "naturalised" as preferred policy options. We then look at *how they work*: how decisions are made and which actors are enrolled in governing processes to lend authority, legitimacy and effectiveness. Next, we discuss *whether, to what extent* and *for whom* they work. We then highlight a series of (un)-intended consequences that flow from these practices and modes of governing that derive from the patterns of delegated authority and forms of valuation upon which they depend. These include accountability and legitimacy deficits, participation gaps and uneven spatial and social development. Finally, we reflect on the ways in which current attachment to these approaches may serve to exclude and neglect adequate consideration of other ways of addressing climate change which may be equally or more effective, but do less to reinforce the power of dominant interests. We conclude by discussing the significance of these observations for debates on climate change, carbon governance and theories of EM.

Ecological Modernisation and the Governance of Carbon

As with many forms of modernism, EM is essentially optimistic about the future. EM theories are interested in the processes of reflexive modernisation that allow societies to recognise and respond to emergent environmental problems (Beck 1995; Christoff 1996; Lash, Szerszynski and Wynne 1996). They are concerned with the discourses through which environment–economy relations are framed and re-framed (Hajer 1995), notably as being mutually supportive rather than antagonistic (Gouldson and Murphy 1996). They have also been used as a basis for prescribing new ways of mediating the relationship between environment and economy (Jänicke et al 1989; Mol 1995; Simonis 1989). Clearly this belief that progress can be made without questioning the basis of modern life has made EM politically attractive, at least in the mainstream.

However, climate change, as a classically "wicked" environmental problem (Rittel and Webber 1973), has challenged the optimism of EM. Such problems are systemic or structural in their origin, operate across multiple scales, occur over long periods of time and will have imperfectly known causes and unexpected consequences (see Berkhout and Gouldson 2003). Capacities first to recognise and then to respond to such problems take some time to emerge. Whilst obviously not uncontested, the accumulation of scientific evidence on climate change over many years has strengthened capacities for problem recognition. But the social response has lagged behind the scientific recognition, partly because of a widespread reluctance to accept the "inconvenient" evidence, and then because of uncertainty about how to respond to a problem where the causes and the consequences are systemic in their nature.

For a period, then, the optimism of the modernists seemed to falter. However, in recent years a substantial degree of technological and economic optimism has reemerged. Technologically, authors such as Pacala and Socolow (2004:968) have claimed that "humanity can solve the carbon and climate problem in the first half of this century simply by scaling up what we already know how to do". Economically, Stern's (2006) claims that the costs of tackling climate change are both relatively affordable and much lower than the costs of not acting have changed the political landscape by making it economically possible and economically necessary to act on climate change. This combination of technological and economic optimism—coupled with claims that tackling climate change could have social, economic, environmental and geo-political co-benefits—has in turn made it politically viable for many governments to set ambitious targets for decarbonisation.

However, this has happened in a context where state capacities to intervene in the economy have become more limited (see Gouldson and Bebbington 2007). Globalisation and liberalisation have meant that the political capital needed to intervene escalates significantly if interventions have negative impacts on economic competitiveness— and such interventions may lead not to the management but merely to the displacement of economic activities through phenomena such as carbon leakage. EM theory would suggest that these limits have led to innovation and policy learning—particularly through a transition away from traditional state-centred forms of intervention and towards more neoliberal and de-centred forms of governance that include a greater emphasis on markets as key delivery mechanisms for environmental governance (Gouldson and Bebbington 2007).

This "governance turn" has thus been associated with a shift away from the "controller" state with its reliance on the hierarchical application of rules and regulations towards the "facilitator" or "enabler" state (Black 2002). Rather than regulating economic activity directly, the facilitator or enabler state seeks to create conditions that allow economic or social actors to govern particular activities. These changes have been reflected in the range of environmental policy instruments that are used in many settings, with experiments with economic and information-based instruments and an increased emphasis on voluntary approaches and different forms of self-regulation. Within these decentred approaches, authority and responsibility are dispersed within broader networks—and so governance becomes a multi-level, multi-actor phenomenon which is complex and fragmented, with new patterns of interaction emerging as a variety of economic and social actors are enlisted to "do the governing" (Black 2002; Bulkeley and Newell 2010).

These developments—the creation of a weight of scientific evidence, the associated increases in social awareness and concern, the gradual reframing of a wicked problem as a technologically, economically and politically tractable problem, experiments with new forms of governance that enable action within the constraints of globalisation and liberalisation—are manifestations of the transformations that are at the heart of processes of EM (Mol and Sonnenfeld 2009). As Hajer (1996) notes, such processes can be interpreted either optimistically as institutional learning—where rational, and responsive institutions learn, adapt and produce meaningful change—or pessimistically as a technocratic project that produces nothing more than a temporary fix to the longer-term structural conflict between environment and economy.

Of these two interpretations, critical perspectives on EM hold that it should be seen as a technocratic project for a number of reasons. Firstly, EM has been criticised for having little to say either on issues of social justice—both in its processual and its distributional forms—or on society–nature relations (Fisher and Freudenberg 2001; Gouldson and Murphy 1996). Secondly, it has been criticised for being spatially contingent, not least because it is based on values and assumptions about institutional capacities that are seen by many to be peculiar to northern Europe (Sonnenfeld and Mol 2006), and because developed countries' attempts at EM often serve to relocate the problems to industrialising and developing countries (Pepper 1998). Thirdly, it has been criticised for creating a hegemony that empowers and legitimises pragmatists who are willing to compromise in their search for politically viable ways forward, whilst simultaneously marginalising and disempowering radicals who hold that such compromises will merely prolong the life of an economic system that needs deeper and more fundamental change if it is to become sustainable (Gibbs 2009). The extent of this hegemony may mean that debates can seem essentially de-politicised since, to be viable, solutions must accept the logic of EM. And fourthly, it has been criticised for focusing on the industrial but not the capitalistic aspects of development, and thereby for overlooking the structural limits of continued economic growth (Mol 2001). Critical perspectives would therefore suggest that the processes of EM would seem to have social, spatial, distributional, political and structural limits.

In the following sections, we apply insights from the conceptual discussion above to two empirical cases: emissions trading and the market for offsets. In each case we consider the political drivers that have led to their emergence and institutionalisation, the ways in which they work and the critical questions relating to whether, to what extent and for whom they work. We conclude by discussing the wider implications of these cases for broader debates on climate change and the governance of carbon and for theories of EM.

Carbon Governance through the EU Emissions Trading Scheme

The EU's 2008 climate and energy package encompasses a range of targets and initiatives on energy efficiency, renewable energy and carbon capture and storage. However, its undoubted flagship is the Emissions Trading Scheme (ETS), a multi-sector and multi-country cap-and-trade scheme which has been the EU's primary instrument for regulating carbon emissions from large energy and industrial installations since 2005. Sectors covered by the scheme account for around 45% of total EU carbon emissions. We begin by examining the drivers for the adoption of this carbon market, and then review its organisation, focusing on interactions between its regulatory and trading dimensions before discussing its effectiveness as a governance mechanism for reducing carbon emissions.

Political Drivers

Significant scholarly attention has been directed towards explaining the EU's move from the international community's major objector to flexibility mechanisms to its leading proponent and practitioner (Bailey 2007; Skjærseth and Wettestad 2008, 2009). Explanations offered include: pressure to show international leadership following the USA's withdrawal from the Kyoto Protocol in 2001 (Schreurs and

Tiberghien 2007); growing maturity of internal debates on emissions trading; dialogue with US officials on the US sulphur dioxide trading scheme; and the influence of advocacy groups like the UK Emissions Trading Group, which had been instrumental in designing the UK's domestic trading scheme. However, Voß (2007) argues that pivotal to these deliberations was the Commission's reframing of emissions trading from a tactical device for diluting binding emissions-reduction commitments to one that promoted effective and cost-efficient action to reduce carbon emissions without jeopardising the competitiveness of European industry.

Thus reframed in accordance with principles of EM, emissions trading compared favourably not just against "command-and-control" regulation (which was seen as expensive and problematic to introduce in the EU context) but also on environmental grounds against its main rival, carbon taxes, because the latter created carbon prices but could not guarantee abatement levels. Emissions trading also circumvented state concerns about ceding tax-raising powers to the EU that had derailed proposals for an EU carbon-energy tax during the 1990s and addressed potential trade distortions that may arise if states adopted different national schemes (Zito 2000). It also appealed to industry concerns about compliance costs and maintaining a level playing field in the Single Market, and to the finance industry, which realised that the trading and banking of emissions allowances provided a potentially lucrative future market (Voß 2007)

Support for emissions trading was not unequivocal, however, especially in Germany and the UK, because of clashes with national measures. Yet one interesting feature of the debate was how few major parties expressed outright opposition to emissions trading. Germany and the UK both accepted a mandatory scheme once they gained concessions for key sectors. Some 2004 accession states, notably Poland, Hungary, Latvia and Lithuania, complained about their inability to influence the scheme prior to their accession but did not oppose its introduction. The European Parliament tabled over 80 amendments to the directive, but again its focus was on securing its preferred outcomes rather than questioning whether the scheme should proceed. Even environmental NGOs like Greenpeace and WWF that originally accused emissions trading of rewarding past pollution ended up hailing it as an important achievement (Ellerman and Buchner 2007).

Organising and Maintaining the EU ETS
Although the broad appeal of emissions trading to EM ideas created critical momentum for the EU ETS, negotiating the details of the scheme exposed multiple tensions between, on the one hand, the need

to standardise the scheme, and on the other, member states' desire to defend their autonomy, competitiveness and the commercial interests of industry sectors affected by the scheme (Bailey and Maresh 2009). The need to maintain the support of the member states and industry groups forced the Commission to accept a relatively decentralised scheme that allowed member states to develop national allocation plans (NAPs) rather than being subject to an EU-wide emissions cap (Kruger, Oates and Pizer 2007; Wettestad 2009a). This gave states significant latitude in how they distributed their Kyoto emissions targets between sectors included and exempted from the scheme and led to the significant overallocation of allowances during the scheme's first phase (2005–2007). This in turn reduced the pressure on firms to cut emissions and triggered a collapse in allowance prices. Another concession allowed the member states to allocate the majority of allowances free of charge rather than by auctioning, giving many energy utilities windfall profits. Other dispensations included exemptions for the aluminium and aviation sectors and the exclusion of the five non-carbon greenhouse gases from the first two trading periods. Finally, the linking directive, which allows target installations to acquire a proportion of their allowances from the Kyoto flexibility mechanisms, provided a further safety valve for companies concerned about the costs of reducing their emissions.

Despite these difficulties, the directive gave the Commission new legal powers to set guidelines, oversee NAPs, and propose reforms to the scheme. More generally, the tribulations of Phase I clarified to all parties the practical actions needed to maintain the market. It also created a degree of policy "lock-in", in that all parties understood that failure to reform the scheme meant defaulting against international commitments and/or the search for a new policy instrument. The Commission began to assert its new powers during Phase II (2008–2012) by requiring revisions to the majority of NAPs because of inconsistencies with Kyoto targets (Brunner 2008). It also initiated a major review of the scheme in 2006, which resulted in proposals for major reforms for Phase III (2013–2020) as part of a revised directive adopted in April 2009.

Despite the unquestionable successes of these strengthening efforts, industry groups and member states continued to lobby for (and gain) concessions, some of which may have significant implications for the scheme in the future (Wettestad 2009b). Among these was the inclusion of provisions enabling energy-intensive sectors that are vulnerable to international competition to continue receiving free allowances during Phase III. A total of 164 sectors—including most steel, non-ferrous metal, cement and oil-refining installations—have since been identified as falling into this category, the free issue of allowances to which may produce a downward effect on prices in the short-to-medium term (European Commission 2009a). The powers of the Commission to shape the NAPs put forward by member states (in particular to specify emissions caps) have also been challenged and at least temporarily curtailed in the European Court of Justice (Court of First Instance 2009).

Alongside the contested but steady strengthening of the EU ETS's regulatory framework, equally important to the regime's maintenance is the monitoring and control of the market itself and the activities of the various target sectors and intermediaries that have joined the market to provide trading services or to speculate on the value of allowances. Here again, aspects of the scheme's design have proven problematic, with a number of questionable market practices emerging. These include:

1 *Carousel fraud:* where bogus traders open accounts in a national registry, buy allowances in another member state without paying VAT (because of exemptions for goods moving between jurisdictions), before selling them on a spot exchange, charging VAT, and disappearing without paying VAT to the authorities.
2 *Phishing fraud:* where installations' identities are stolen from online registries to gain access to allowances.
3 *Gaming practices:* these include *strip-and-swap* deals, where traders swap EU allowances for the Kyoto Protocol's Clean Development Mechanism (CDM) credits, giving clients additional profits from price differentials between the credits and themselves additional commission—and *fuel switching*, where large energy utilities buy allowances and announce a switch to a high-carbon energy source to profit from increases in allowance prices caused by the announcement.

Convery (2009) argues that such practices have a minimal effect on the wider market and that smoking guns like windfall profits from over-allocations provide good opportunities to strengthen the scheme. VAT fraud alone is nevertheless reported to account for around €5 billion annually, and prompted the Commission to submit urgent proposals to address the problem (Ainsworth 2009). More broadly, the cases indicate the multitude of opportunities that exist for market actors to boost profits within a shadow economy running in parallel to, but with different objectives to, mainstream carbon trading.

Environmental and Political Effects of the EU ETS
Unlike the broader social and environmental goals of carbon-offset mechanisms like the CDM, the primary purpose of the EU ETS is to reduce carbon emissions from power producers and energy-intensive industries. Accordingly, this section focuses on the emissions reductions achieved so far by the scheme and on its wider political implications.

This is not to make light of the social and distributional issues that underpinned many disputes during the EU ETS negotiations, but reflects the fact that emissions reduction is the ultimate benchmark against which the success or failure of the EU ETS will be decided.

While the full effects of the EU ETS on emissions will only be revealed as allowance scarcity increases during Phases II and III, Phase I of the scheme was widely recognised as being ineffectual as a result of the over-allocation of allowances. However, a 3.06% reduction in emissions among ETS sectors was recorded between 2007 and 2008. The European Commission (2009b) noted that although this was partly caused by the economic slowdown, it also claimed that the fall reflected measures undertaken in response to the price signal created by the scheme and was significant in the context of 0.8% GDP growth. A further 11.2% fall was recorded in 2009, with reductions exceeding 30% in some sectors (Reuters 2010). However, the data also show a 3.6% over-allocation of free allowances compared with actual emissions across the scheme for 2009, suggesting that recessionary effects rather than caps have so far been the main driver of emissions cuts.

However one interprets these data, the EU ETS has had a transformative effect on climate politics within the EU. The EU's central task in the aftermath of the Kyoto negotiations was to find a practical strategy to show that EU leadership on climate change would not be economically ruinous in the absence of commitments by other major countries. By reframing emissions trading from a tactical device to dilute emissions commitments to an effective and efficient instrument for reconciling environmental protection with economic growth, the Commission was able to build a strategic coalition for the creation of an EU climate-governance regime (Voß 2007).

While this reframing of emissions trading as a practical device for realising the ambitions of EM could be considered as a major breakthrough, our account also reveals the immense difficulties the EU has encountered in creating and maintaining the EU ETS. At the heart of these difficulties is a fundamental tension between the regulatory and territorial logics of emissions trading (Bailey and Maresh 2009); between the willingness of key actors to subscribe, in principle, to the idea that there are technologically and economically viable responses to climate change, and to accept, in practice, the political, economic and geographical realities of devising a market-based carbon regime.

Each stage of the regime-building process has required mediation between competing state and industry interests, and the Commission's ingenuity in strengthening the scheme itself testifies to the learning capacities of EM. At the same time, these mediation efforts have led to a raft of unforeseen consequences that hindered the effectiveness of the EU ETS during its early years. Such problems are not exclusive to market-based forms of environmental governance (Kruger, Oates and Pizer 2007); however, the dispersal of agency inherent in emissions trading, where everyone has some control but no-one has full control or accountability for achieving policy goals, has created a multitude of opportunities for target sectors and intermediaries to pursue rent-seeking behaviour that has compounded the unintended outcomes produced by the EU ETS negotiations.

What is perhaps equally striking, however, is the restricted nature of debates among those at the core of EU climate policy on the appropriateness of emissions trading as a lynchpin of the EU's climate strategy (Castree 2010). Those that have taken place have been dominated by the same political and economic elites that created the scheme and have focused mainly on the mechanics and distributional economics of emissions trading and on positive framings of market-based instruments. These foci have limited the opportunity for alternative ways of reducing energy and industrial emissions—such as greater direct investment or focusing on behavioural changes—to gain real traction in mainstream EU policy (Bailey and Wilson 2009). The growing interest of other countries (especially the USA) in emissions trading as a means of regulating energy and industrial emissions as a result of the EU ETS has had a similar effect at the international level (Skjærseth and Wettestad 2008).

Carbon Governance through the Market for Offsets

The development of carbon offset markets has emerged as an increasingly critical component of climate governance in general, and the carbon economy in particular. We have seen the rapid development of offsets overseen by the CDM within what is known as the compliance market alongside the spectacular growth of a voluntary market in offsets for companies and individuals. While the economic slump dampened activity in 2008, the global value of primary offset transactions had grown to US$7.2 billion in 2008, representing a more than 10-fold growth from 2004, largely due to the CDM. The CDM accounts for 90% of offset-project transaction volumes and value, and therefore provides the main focus here. Joint Implementation (JI), the mechanism supporting offset projects within developed-country parties to the Kyoto Protocol, accounts for another 5% and voluntary offsets for the remaining 5%. In contrast to declines in primary transactions for both JI and CDM, the voluntary market continued to post double-digit growth in 2008 (Capoor and Ambrosi 2009). The ability of both compliance and voluntary offset markets to continue on this trajectory will depend on their ability to enrol the support of powerful government and industry backers and to manage criticisms from carbon traders and project developers about their efficiency and governance and from opponents of carbon markets who question their credibility and effectiveness.

While both markets have distinct origins and history, they raise similar governance challenges and engage many of the same actors in overlapping policy networks. What is interesting politically, and in terms of EM, is that a political constituency around offsets has been created. This includes traders, intermediaries, project developers and accountants who have an interest in safeguarding the credibility of offset markets and seeing them succeed over other ways of responding to climate change which are perceived to be less business friendly (Newell and Paterson 2010). This is important to understanding the role of specific issue-based and policy networks in sustaining wider coalitions that seek to protect their political influence and market share and, ultimately, the legitimacy of neoliberal responses to climate change.

Political Drivers

The history and evolution of these markets is revealing of their dynamics and relationship to neoliberalism. The CDM has been described as the "Kyoto surprise" because of the rapid and seemingly haphazard way in which it emerged as a product of an 11th-hour negotiation at Kyoto in 1997 (Werksman 1998). Very much a creation of political necessity drawing on Brazilian proposals concerning a Clean Development Fund, its details were worked out in informal contact groups in the final days of Kyoto, spearheaded by the Brazilian delegation with US support. Much of the detail of how it would work was left to the meeting in Marrakesh in 2001, where accords were approved on the rules and modalities of its operation.

At a deeper level, the political and commercial drivers for flexible market-based approaches, particularly mechanisms like the CDM which opened up a global market in lowest-cost emissions reductions, included the desire to avoid imposing costs on powerful nations and sectors that feared loss of competitive advantage if emissions cuts were required of them but not of their emerging competitors in countries such as India and China. This logic underpinned the US's rejection of the Kyoto Protocol supported by an aggressive lobbying campaign by many US companies and was embodied in the Byrd-Hagel Senate resolution that prevented the USA from ratifying a treaty that did not include binding emissions cuts for leading developing countries.

The voluntary market meanwhile was set up to meet a growing demand by businesses and individuals concerned about their carbon footprint. It tapped into the rise of climate change as an issue of corporate social responsibility and the pressure on firms to account for and reduce their emissions even if through offset purchases rather than their own abatement efforts (Begg, van der Woerd and Levy 2005). Ridiculed by activists as being equivalent to the "indulgences" of the middle ages, when the rich could pay poorer people to go to prison on their behalf (Smith 2007), the creation of these markets has led to attempts to differentiate offsets which deliver cuts *and* social benefits from those which do neither. Indeed, various private standards and certification schemes have been created to address some of the concerns that greeted the first wave of offset projects (Newell and Paterson 2010).

Organising and Maintaining Offset Markets

The way offset markets have been organised encapsulates the EM logic of reconciling growth and environmental protection in that innovators and project developers are emboldened and enabled to search for the least-cost way of producing emissions reductions irrespective of the ethical and political dilemmas around, for example, emissions necessary for survival as opposed to luxury consumption around which critics of carbon markets draw clear distinctions (Agarwal 2000). They rest on the premise that a tonne of carbon saved is equivalent wherever the reduction occurs.

The modes of governance that have emerged around offset markets are also indicative of neoliberal modes of environmental governance in general. These include the nature of the actors involved and the ways they work together—often through voluntary public–private partnerships or hybrid arrangements (Bäckstrand 2008; Streck 2004), the creation of networks for the pursuit of particular policy goals in preference over a dominant role for the state, and a basic ideological commitment to market approaches and the efficiency they are claimed to deliver (Newell and Paterson 2009). They include governance through accountancy, disclosure and audit, all necessary for the creation of a fungible and commensurate unit that can be commodified and traded and for generating trust among investors that they are buying a credible product.

In the case of offsets the boundaries of the policy network are perhaps more porous, desegregated and multi-scalar than those for the EU ETS. CDM governance spills over into, and is simultaneously defined by, other policy areas at all levels of decision-making because it covers sectors as diverse as energy, agriculture and forestry. It touches, therefore, on entrenched and powerful regimes of governance in other areas with their own actor-networks, conflicts of interests and programmes of regulation that pre-date the CDM and seek to engage it on their own terms by trading access to projects for a share of carbon finance (Newell 2009).

The breadth and range of networks of actors that have to be enrolled to make the offsets market function creates a series of mutual dependencies and requires the exchange of political and other resources. The Conference and Meetings of the Parties ultimately exercises authority over the remit of the CDM. The CDM Executive Board, which reports to the Conference and Meetings of the Parties, approves methodologies and accredits Designated Operational Entities (DoEs) who are conferred the authority to assess whether projects have delivered claimed emissions reductions and therefore should be issued with Certified Emissions Reductions (CERs). With just 10 members, the CDM Executive Board has to delegate significant power and authority to DoEs to approve projects and then later to verify that they have achieved the claimed emissions reductions. Given there are just 18 approved DoEs and far fewer that are approved to operate in each individual sector in which the CDM is active, DoEs inevitably end up having to approve one another's projects. This potentially creates openings for collusion and disincentives to criticise or reject another firm's project for fear they will reciprocate (Green 2008). Nevertheless, to maintain the environmental credibility on which the system notionally rests, sanctions occasionally need to be applied. Indeed several DoEs have had their accreditation suspended for continually proposing projects that do not fulfil the minimum approval criteria. There have also been calls to address potential conflicts of interest between members of the CDM Executive Board, the roster of experts upon which they call for guidance, and the projects they have been involved in developing but are then charged with assessing.

Apart from the multiple scales at which offset markets operate and have to be governed, there are also issues of how the markets, and the policy networks which create and maintain them, interconnect. The CDM is connected to other carbon markets, such as the EU ETS, through the linking directive which is a significant source of demand for the purchase of CERs. Likewise, the Chicago Climate Exchange, which allows those companies that have committed to a voluntary emissions reduction to buy and sell carbon credits, was often the final destination for rejected CDM projects but for which demand existed on the voluntary market. It is not yet clear the extent to which a globally integrated emissions trading regime is possible, but carbon is increasingly passing through and simultaneously creating global circuits of capital. While opening up new reduction and accumulation opportunities, their connection to distant markets and dependence upon the networks which govern them to exercise quality control, over which governments or market actors often have very little control, exposes them to the political fallout that may ensue from the purchase of "sub-prime carbon" (Friends of the Earth 2009).

On the demand side, market actors need strong signals from climate negotiators to create conditions of scarcity and to drive demand for the products they are selling. They also require public institutions to develop rules of conduct for predictability and credibility. At the same time, businesses often complain about the overly stringent application of additionality criteria and of delays in the approval process. They want maximum flexibility and a more harmonised and scaled-up process including further moves towards programmatic or sectoral CDM, which open up many more opportunities at lower cost. There are divisions among traders between those anxious to safeguard the environmental integrity of the market on which their profitability rests and those with less at stake in terms of long-term reputation that want to maximise short-term profits. Dubbed "cowboy capitalists", these actors have tended to be concentrated in the voluntary market and have prompted efforts to provide buyers of offsets with firmer guarantees of their credibility through initiatives like the Offset Quality Initiative. The balance of power among and between these actors who share common political ground but competing commercial interests and preferences continues to shape the governance of offsets.

The shadow of regulation and intervention by public authorities, as well as criticism and negative media exposure, creates an incentive for new forms of private and voluntary regulation to protect the credibility of the market as a whole and to preserve the autonomy of the network of traders and brokers to set their own forms of *modus operandi*. It requires them to address the governance problems that arise in relation to quality control, integrity, independence and credibility. For example, the "Climate, Community and Biodiversity" standards aim to deliver benefits to host communities and help protect biodiversity where a strict emphasis on greenhouse gas reductions may not always guarantee this. Each of the initiatives that has been set up to address specific concerns about the integrity of offsets and their beneficiaries entails enrolling a broader network of actors in order to lend them credibility and ensure their smooth operation.

The Environmental and Political Effects of Offset Markets

The organisation of offset markets clearly raises questions about who wins and who loses from such responses to climate change. Just as offset markets reflect other features of neoliberalism, they also embody its tendencies towards uneven development and the (re)production of inequality on the one hand and problems of corruption and fraud on the other.

In relation to uneven development, despite the hopes that carbon finance would generate new revenue streams for the poor, capital flows in the CDM have largely mirrored flows of foreign direct investment in the developing world, with China, India and Brazil the three largest recipients while sub-Saharan Africa continues to attract less than 2% of CDM projects. Efforts have also concentrated on the search for low-hanging fruit or the largest volume of emissions reductions for the lowest possible investment. In the CDM market, 70% of CERs in the first 18 months were issued for abating gases other than carbon dioxide, in particular the destruction of industrial gases used in refrigeration (Paulsson 2009). Because of the fact CERs are weighted according to the global warming potential of a gas, this creates incentives to target cheap and easy projects aimed at removing gases such as hydrofluorocarbons rather than more difficult investments in renewables for example.

Moreover, the drive to capture value at minimum cost makes it rational to make minor changes to existing production processes in order to get credit (and climate finance) for emissions reductions. In the worst case, this can be a subsidy to polluting activities which affect the poor most seriously, where, for example, waste sites are given a new lease of life by receiving payment for capturing and burning methane, undermining campaigns for their closure (Lohmann 2005). The ways in which value is apportioned also explains the lack of attention to the sustainable development benefits of projects, in terms of employment and sourcing of local materials for example, which forms the second rationale for and requirement of CDM projects, and was a condition for the support of developing countries for the CDM. While CERs are awarded for quantifiable emissions reductions, the same is not true for contributions to sustainable development. To capture value associated with those contributions, other forms of private regulation have been developed in carbon markets such as the CDM Gold Standard or in relation to forestry projects the Climate, Community and Biodiversity standards and Plan Vivo initiative.

In relation to the claim that carbon markets entrench capitalism's tendency to generate corruption and fraud, activists have exposed "climate fraud" where the same CDM project is sold to multiple buyers and double-counted, or many reported cases of failure to meet basic additionality criteria (Bachram 2004; Lohmann 2005). This requires demonstrating beyond reasonable doubt that a specified volume of emissions would not have taken place without the investment and technology associated with a CDM project.

In sum, while the high level of demand for CDM projects and the rate at which the market has expanded has been a (welcome) surprise to many, the future of the CDM remains in doubt because its fate is tied to that of the Kyoto Protocol. It is also the case that technology transfer has largely not been achieved, sustainable development benefits have not been prioritised and, as with neoliberalism in general, there has been a tendency towards a concentration of capital in the wealthier parts of the world.

Concluding Discussion

Over the last decade the emerging global trade in carbon has become increasingly central to efforts to govern climate change. As Newell and Paterson argue (2009:80) "Climate politics is increasingly conducted by, through and for markets". The significance of existing approaches to emissions trading and the market for offsets, and the tide of political opinion favouring the further expansion of carbon markets, underscores the need for critical examination of the ideological foundations and social, environmental and spatial consequences of the new carbon economy.

The cases explored in the previous two sections—emissions trading schemes and markets for offsets—exemplify how new market-based forms of carbon governance have emerged and are evolving. Both show that a major force driving carbon markets has been their appeal to a broad community of state and non-state actors. For international bodies and governments, they offered a way of meeting commitments to cut greenhouse gases without sacrificing other policy goals; for emitting industries, they provided a "least-worst" option for managing the risks of regulation and the commercial risks associated with climate change; and for market intermediaries and speculators, they have created new commercial opportunities. By winning the support of this range of actors, there does seem to have been a process of institutional learning that has rendered a wicked environmental problem more tractable.

But limited evidence exists yet of a consensus on the depth and types of reform that such market-based forms of governance should deliver. The main rifts within the EU ETS have been between the Commission's prioritisation of international commitments and the member states and industry groups' reluctance to pursue decarbonisation except where it does not threaten established economic-commercial interests. Similar diversity is evident in carbon offset markets, where there is consensus on the desirability of offset schemes but rising discord about their management. This was clearly evident in heated exchanges at meetings in Copenhagen between the CDM Executive Board's attempts to remain true to the mandate conferred upon it by the climate regime in terms of guaranteeing "additional" emissions reductions and growing frustration among DoEs and project developers at alleged inconsistent applications of rules in the CDM market which impact their profit margins. The consensus that underpins these new forms of carbon governance may therefore depend both on their ability to maintain buy-in from a wide range of actors and their inability to challenge the interests of those actors. These new forms of governance may well be efficient at channelling resources towards the easiest options for decarbonisation for a period, and, notwithstanding the issues of corruption and fraud discussed above, they may be effective at delivering carbon emissions reduction whilst doing so. But once the easy options have dried up it is far from clear whether the consensus that underpins them and the governance arrangements that manage them are strong enough to deliver the deep and rapid cuts in carbon emissions that climate science suggests are necessary. It may be that these new governance arrangements can only deliver the targets that are agreed upon by more traditional, state-centred, forms of government.

Another shared feature of the two carbon markets is the multi-scalar, multi-actor nature of the networks involved in their oversight and operation. Actors are generally enrolled on the basis of their expertise and capability to make carbon markets function rather than according to criteria of democratic representation or equity, and networks are normally concentrated in centres of political and financial power. Such a process of network formation with spatial roots but international reach has important spatial and social justice implications. It effectively excludes the majority of developed and developing country populations whose lives are affected by climate change *and* by carbon markets from participating in key decisions about what these markets should be achieving, where, and by what means. It also excludes those critical voices that do not have the political or economic resources needed to gain entry to the core policy networks and arenas. Whilst these new forms of carbon governance may offer some (perhaps temporary) benefits in terms of tractability, efficacy and efficiency, these gains may be realised at the expense of equity, legitimacy and accountability.

A final issue concerns the degree to which the recent formation of these market-based approaches has locked corporations, countries and the wider international community into a neoliberal experiment with climate governance. The EU ETS has frequently been described as a grand policy experiment (Wettestad 2005); certainly, the investments made—both financial and political—make abandoning the scheme prior to 2020 at the earliest virtually unthinkable. Similar political and economic reserves have been expended on the CDM and proposed reforms reflect its rapid growth which has created capacity bottlenecks that need to be addressed. The tide of political opinion instead appears to be running in favour of further strengthening of ties through the linking of national and regional emissions trading schemes (Grubb 2009), in effect further globalisation and locking in of this neoliberal/EM experiment and all its social and spatial implications and structural weaknesses. But in a context where market-based approaches have been institutionalised and legitimised, albeit conditionally, the opportunity for more radical voices to impact upon the dominant framings or governance processes seems very limited indeed.

These critiques should not be seen necessarily as an argument against carbon markets or valuing environmental resources in principle. Clearly, under-pricing of nature's services has been a major cause of their over-exploitation, and alternative proposals such as carbon taxes have their own sets of governance issues, not least concerns over sovereignty where the regional or global collection of taxation revenues is proposed. They also have to address carbon leakage in a globalised and integrated economy and are just as susceptible to special-interest lobbying as emissions trading and offset schemes. However, even if one sets aside the structural problems of EM as an ideological and practical framing of how climate change should be managed, our review of emissions-trading and carbon-offset markets reveals a multitude of social and spatial challenges that those involved in governing carbon markets must address in order to maintain their credibility.

In policy terms, this is likely to require more energetic regulation of emissions caps, the criteria used for issuing carbon credits, and how schemes are linked. Key issues also relate to the transparency, accountability and legitimacy of both the processes and the outcomes of these experiments with carbon governance (Transparency International 2011). Closer scrutiny of actor behaviour within carbon markets will also be needed to deter more obvious market abuses, and public and private actors involved in carbon governance will have to find new ways of arbitrating between competing sectoral and geographical interests where uneven exposure to competition and uneven access to carbon markets inhibit the ambitions of carbon market enthusiasts. Carbon markets gained much of their prominence during a period of apparently unshakable faith in markets to deliver economic prosperity, environmental protection and social provisions. It is an interesting question whether this approach would have been embraced as unequivocally in a post credit-crunch world, especially if its leading advocate, the USA, remained absent from international climate treaties. It might also be a rather academic one were it not for the path-dependency of existing and planned new carbon markets. The drift towards more directive state involvement (albeit in a context where state capacities seem more limited) in climate and energy policy suggests that a more varied set of measures combining markets, regulations, and investment programmes is likely.

With regard to EM theories, in our conceptual discussion we noted Hajer's (1996) suggestion that EM can be interpreted either as institutional learning or as a technocratic project. Much of the analysis of new forms of market-based carbon governance in this paper reinforces the critical view that EM should be seen as some of the former but much of the latter. Market-based forms of carbon governance do seem to have been based on a process of institutional learning that has rendered a wicked environmental problem more tractable. But they also display many features of a technocratic project: they focus on efficacy and efficiency but have little to say on issues of social justice; they seem to be spatially contingent and to generate uneven patterns of spatial development; and they appear to reinforce a neoliberal hegemony that empowers those pragmatists who are willing to engage whilst simultaneously disempowering its critics and excluding approaches that could lead to deeper change.

We began this paper by suggesting that EM theories are centrally concerned with social capacities to recognise and respond to existing and emergent environmental problems. We conclude by questioning whether we currently have the social capacities to recognise and respond to the emergent limits of these new forms of carbon governance in time to avert dangerous climate change.

Acknowledgments

The authors thank participants at the Carbon Economy workshop, hosted by the Environmental Change Institute, Oxford University, in October 2009, for their many insightful comments. Andy Gouldson acknowledges the support of the ESRC *Centre for Climate Change Economics and Policy*; Peter Newell financial assistance from the ESRC *Climate Change Leadership Fellowship* programme; and Ian Bailey funding by the British Academy (Ref: SG-54196) for the project *Political Strategies for Future Climate Policy*.

References

Agarwal A (2000) Global warming in an unequal world. *Equity Watch*, 15 November, http://www.cseindia.org/campaign/ew/art20001115_2.htm (last accessed 20 December 2009)

Ainsworth R (2009) The morphing of MTIC fraud: VAT fraud infects tradable CO_2 permits. Boston University School of Law Working Paper No 09–35, http://papers.ssrn.com/sol3/papers.cfm?abstract_id=1443279 (last accessed 21 April 2010)

Bachram H (2004) Climate fraud and carbon colonialism: The new trade in greenhouse gases. *Capitalism, Nature, Socialism* 15(4):10–12

Bäckstrand K (2008) Accountability of networked climate governance: The rise of transnational climate partnerships. *Global Environmental Politics* 8(3):74–102

Bailey I (2007) Neoliberalism, climate governance and the scalar politics of EU emissions trading. *Area* 34(1):431–442

Bailey I and Maresh S (2009) Scales and networks of neoliberal climate governance: The regulatory and territorial logics of European Union emissions trading. *Transactions of the Institute of British Geographers* 34(4):445–461

Bailey I and Wilson G (2009) Theorising transitional pathways in response to climate change: Technocentrism, ecocentrism, and the carbon economy. *Environment and Planning A* 41(10):2324–2341

Beck U (1995) *Ecological Politics in an Age of Risk*. Cambridge: Polity

Begg K F, van der Woerd F and Levy D (eds) (2005) *The Business of Climate Change: Corporate Responses to Kyoto*. Sheffield: Greenleaf Publishers

Berkhout F and Gouldson A (2003) Shaping, modulating, adapting: Perspectives on technology, environment and policy. In F Berkhout, M Leach and I Scoones (eds) *Negotiating Change: Advances in Environmental Social Science* (pp 231–260). Cheltenham: Edward Elgar

Black J (2002) Critical reflections on regulation. *Australian Journal of Legal Philosophy* 27(1):1–35

Brunner S (2008) Understanding policy change: Multiple streams and emissions trading in Germany. *Global Environmental Change* 18(3):501–507

Bulkeley H and Newell P (2010) *Governing Climate Change*. London: Routledge

Capoor K and Ambrosi P (2009) *State and Trends of the Carbon Market 2008*. Washington DC: World Bank

Castree N (2010) Crisis, continuity and change: Neoliberalism, the left and the future of capitalism. *Antipode* 42(1):1327–1355

Christoff P (1996) Ecological modernization, ecological modernities. *Environmental Politics* 5(3):476–500

Convery F (2009) The emerging literature on emissions trading in Europe. *Review of Environmental Economics and Policy* 3(1):121–137

Court of First Instance (2009) Judgments of the Court of First Instance in Case T-183/07 and Case T-263/07 Poland v Commission, Estonia v Commission. Press Release No 76/09, 23 September 2009

Ellerman A and Buchner B (2007) The European Union emissions trading scheme: Origins, allocation, and early results. *Review of Environmental Economics and Policy* 1(1):66–87

European Commission (2009a) Draft Commission Decision of determining, pursuant to Directive 2003/87/EC of the European Parliament and of the Council, a list of sectors and subsectors which are deemed to be exposed to a significant risk of carbon leakage. C(2009) xxx2009. Brussels: European Commission

European Commission (2009b) Emissions trading: EU ETS emissions fall 3% in 2008. *Europa Press Release* RAPID IP/09/794, 15 May

Fisher D and Freudenberg W (2001) Ecological modernisation and its critics: Assessing the past and looking toward the future. *Society and Natural Resources*, 14(8):701–709

Friends of the Earth (2009) *Sub-Prime Carbon: Re-thinking the World's Largest New Derivatives Market*. Washington: Friends of the Earth

Gibbs D (2009) *Climate Change, Resource Pressures and the Future for Regional Development Policies*. Northern Way Turning Points Paper, Centre for Urban and Regional Development Studies. Newcastle: University of Newcastle

Giddens A (2009) *The Politics of Climate Change*. Cambridge: Polity Press

Gouldson A and Bebbington J (2007) Corporations and the governance of environmental risks. *Environment and Planning C* 25(1):4–20

Gouldson A, Hills P and Welford R (2008) Ecological modernisation and policy learning in Hong Kong. *Geoforum* 39(1):319–330

Gouldson A and Murphy J (1996) Ecological modernisation and the European Union. *Geoforum* 27(1):11–21

Green J (2008) Delegation and accountability in the Clean Development Mechanism: The new authority of non-state actors. *Journal of International Law and International Relations* 4(2):21–55

Grubb M (2009) Linking emissions trading schemes. *Climate Policy* 9(4):339–340

Hajer M (1995) *The Politics of Environmental Discourse: Ecological Modernisation and the Policy Process*. Oxford: Clarendon Press

Hajer M (1996) Ecological modernization as cultural politics. In S Lash, B Szerszynski and B Wynne (eds) *Risk, Environment and Modernity: Towards a New Ecology* (pp 246–268). London: Sage

Jänicke M, Monch H, Rannenberg T and Simonis U (1989) Structural change and environmental impact. *Environmental Monitoring and Assessment* 12(2):99–114

Kruger J, Oates W E and Pizer W (2007) Decentralization in the EU emissions trading scheme and lessons for global policy. *Review of Environmental Economics and Policy* 1(1):112–133

Lash S, Szerszynski B and Wynne B (eds) (1996) *Risk, Environment and Modernity: Towards a New Ecology*. London: Sage

Lohmann L (2005) Marketing and making carbon dumps: Commodification, calculation and counterfactuals in climate change mitigation. *Science as Culture* 14(3):203–235

Mol A (1995) *The Refinement of Production: Ecological Modernization and the Chemical Industry*. Utrecht: Van Arkel Publishers

Mol A (2001) *Globalization and Environmental Reform: The Ecological Modernization of the Global Economy*. Cambridge: The MIT Press

Mol A and Sonnenfeld D (2000) Ecological modernisation around the world: An introduction. *Environmental Politics* 9(1):3–14

Mol A, Sonnenfeld D and Spaargaren G (eds) (2009) *The Ecological Modernization Reader: Environmental Reform in Theory and Practice*. London: Routledge

Murphy J and Gouldson A (2000) Integrating environment and economy through ecological modernisation? An assessment of the impact of environmental policy on industrial innovation. *Geoforum* 31(1):33–44

Newell P (2009) Varieties of CDM governance: Some reflections. *Journal of Environment and Development* 18(4):425–435

Newell P and Paterson M (2009) The politics of the carbon economy. In M Boykoff (ed) *The Politics of Climate Change: A Survey* (pp 80–99). London: Routledge

Newell P and Paterson M (2010) *Climate Capitalism*. Cambridge: Cambridge University Press

Pacala S and Socolow R (2004) Stabilization wedges: Solving the climate problem for the next 50 years with current technologies. *Science* 305(5686):968–972

Paulsson E (2009) A review of the CDM literature: From fine-tuning to critical scrutiny? *International Environmental Agreements: Politics, Law and Economics* 9(1):63–80

Pepper D (1998) Sustainable development and ecological modernization: A radical homocentric perspective. *Sustainable Development* 6(1):1–7

Reuters (2010) EU Emissions Trading Scheme 2009 data. http://uk.reuters.com/article/idUKLDE63010120100401 (last accessed 12 May 2010)

Rittel H and Webber M (1973) Dilemmas in a general theory of planning. *Policy Sciences* 4(2):155–169

Schreurs M and Tiberghien Y (2007) Multi-level enforcement: Explaining European Union leadership on climate change mitigation. *Global Environmental Politics* 7(4):19–46

Simonis U (1989) Ecological modernization of industrial society: Three strategic elements. *International Social Science Journal* 41(3):347–361

Skjærseth J and Wettestad J (2008) *EU Emissions Trading: Initiation, Decision-Making and Implementation*. Aldershot: Ashgate

Skjærseth J and Wettestad J (2009) The origin, evolution and consequences of the EU emissions trading system. *Global Environmental Politics* 9(2):101–123

Smith K (2007) *The Carbon Neutral Myth: Offset Indulgences for your Climate Sins*. Amsterdam: Carbon Trade Watch

Sonnenfeld D and Mol A (2006) Environmental reform in Asia. *Journal of Environment & Development* 15(2):112–137

Stern N (2006) *The Economics of Climate Change*. Cambridge: Cambridge University Press

Streck C (2004) New partnerships in global environmental policy: The Clean Development Mechanism. *Journal of Environment and Development* 13(3):295–322

Transparency International (2011) Global Corruption Report: Climate Change. London: Earthscan.

Voß J-P (2007) Innovation processes in governance: The development of emissions trading as a new policy instrument. *Science and Public Policy* 34(5):329–343

Werksman J (1998) The Clean Development Mechanism: Unwrapping the Kyoto surprise. *Review of European Community and International Environmental Law* 7(2):147–158

Wettestad J (2005) The making of the 2003 EU emissions trading directive: An ultra-quick process due to entrepreneurial proficiency? *Global Environmental Politics* 5(1):1–23

Wettestad J (2009a) European climate policy: Toward centralized governance? *Review of Policy Research* 26(3):311–328

Wettestad J (2009b) EU energy-intensive industries and emission trading: Losers becoming winners? *Environmental Policy and Governance* 19(5):309–320

Zito A (2000) *Creating Environmental Policy in the European Union*. London: Macmillan

Chapter 6
Accounting for Carbon: The Role of Accounting Professional Organisations in Governing Climate Change

Heather Lovell and Donald MacKenzie

Introduction

Climate change continues to be an important issue on national and international policy agendas (DEFRA 2007; IPCC 2007). To date the most prominent way the problem of climate change has been addressed is through the construction of markets in which standard units of greenhouse gas emissions are created and exchanged. A fuller understanding of both the potential and the weaknesses of carbon markets requires not just economics (the source of nearly all existing work on them) but also investigation of the implications of carbon markets for other disciplines and professional activities: interdisciplinary work on accounting is crucial in this respect, and to date has been somewhat overlooked by researchers in fields other than accounting (see MacKenzie 2006 for an exception). Practices of many kinds are needed to successfully commoditise carbon and make carbon markets work and, amongst these, accounting is of particular importance. In this paper therefore we assess the role of the accountancy profession in governing the new carbon economy, focusing on the role of the main international accountancy professional organisations and the work they do in appointing accountants as managers of carbon.

We seek to answer key empirical questions about the governance of climate change by accountants: how has the accountancy profession been involved in the day-to-day governance of climate change to date?

The New Carbon Economy, First Edition. Edited by Peter Newell, Max Boykoff and Emily Boyd.
© 2012 Heather Lovell and Donald MacKenzie. Book compilation © 2012 Editorial Board of Antipode and Blackwell Publishing Ltd.

What is its significance? How has the authority of accountants as carbon managers been established? These questions are relevant in furthering our conceptual understanding of the new political and institutional challenges that flow from managing the new carbon economy, and in particular the new alliances, coalitions, resistances that are emerging aimed at engaging, embedding or rejecting it. These questions are also significant for the operation of carbon markets, since accounting makes economic items visible, and whether and how it does so is consequential. In a relatively new area of policy such as climate change where societal responses are still in flux there is an excellent and valuable opportunity to better understand accounting decision-making processes before they become embedded and routinised.

The paper concentrates primarily on financial accounting (accounting for carbon in financial accounts, and the activities of financial accounting professional bodies and standard setters in relation to climate change). This is because financial accounting is a central means by which firms in a capitalist society report on their activities. Where relevant, however, we also discuss the role of auditing and management accountancy. There is a somewhat "grey" area between carbon financial accounting and non-financial (so-called "narrative") disclosure of corporate climate impact and carbon benchmarking; it is the latter area of activities indeed where the term "carbon accounting" has recently become most prevalent (see for example The Aldersgate Group 2007). For this reason debates and activities at the intersection of corporate reporting and financial disclosure (eg the work of the Climate Disclosure Standards Board) are relevant and considered here. Mainly, however, we concentrate on the role of financial accounting professional organisations, especially those such as the Canadian Institute of Chartered Accountants (CICA) and the Institute of Chartered Accountants in England and Wales (ICAEW) who have taken the lead in engaging with climate change. We draw on in-depth interviews (20 in total) with key industry players active in carbon accounting, including accountancy firms, standard setters (the IASB or International Accounting Standards Board, and the main US body, the Financial Accounting Standards Board, FASB), and financial accountants at large European companies active in emissions trading. These interviews have been transcribed and coded.[1] The paper is based on research funded by the UK Nuffield Foundation and is part of a wider project investigating precisely how carbon is being made fungible (ie standardised and interchangeable), using ideas from economic sociology and political science.[2] The paper builds on initial exploration of accounting for carbon in the EU Emissions Trading Scheme (EU ETS), investigating in more depth the preliminary findings and research themes identified by MacKenzie (2009). The role of accounting professional organisations in climate change governance was a finding that emerged somewhat unexpectedly out of interviews, which were focused at the outset more narrowly on EU ETS financial accounting practices.

The paper also draws on research conducted by Lovell and others,[3] funded by the Association of Certified Chartered Accountants (ACCA) and the International Emissions Trading Association (IETA) involving a survey of carbon financial accounting practices of top emitters in the EU ETS. Lovell also is a member of the Climate Disclosure Standards Board (CDSB) Technical Working Group, and uses some data and ideas from her role with the CDSB in this paper.

The theoretical frameworks judged to be most relevant and illuminating in relation to exploring the response of accountants to climate change centre on issues of measurement, calculation, and expertise, and are drawn variously from accountancy and society literatures, covering the history of accountancy and critical examination of the practices and culture of accountancy (Hopwood and Miller 1994; Miller 1994), Foucault's theory of governmentality (Dean 1999; Foucault 1991), and the policy network concept of "epistemic communities" (Haas 1992a; 1992b). These diverse literatures are especially helpful in thinking about how authority is gained through promoting uptake of certain seemingly neutral practices and techniques (eg the application of financial accounting principles and techniques to climate change—double-entry book keeping; quantitative and narrative formats etc), and through discourse (eg the discursive positioning of accountancy as the "natural home" for the professional management of carbon). Concepts and ideas from these literatures are used as lenses to examine more widely the political and institutional challenges of governing carbon for accountants and the accountancy profession.

The paper is structured as follows. First, relevant literatures are reviewed—as discussed—to interrogate the key governance issues for accountants and climate change. Second, in the main empirical section of the paper we give a short history of the involvement of accountants with climate change ("Stage One"—late 1990s to 2005) and then examine how and why accountancy professional organisations have more recently attempted to position themselves and the profession as well placed to govern climate change ("Stage Two"—2005 to present). In conclusion, we comment on the likely future directions of the carbon accounting debate and its implications for policy and theory.

How and Why Carbon is Measured

Conceptualising carbon accounting potentially cuts across a number of different theories and bodies of research. There is a range of relevant literatures to draw upon which offer useful insights into how and why accountants might be framing themselves as good and "rightful" managers of carbon. Here we briefly consider three literatures judged to be most relevant. The first body of work—broadly termed "society and accountancy"—examines issues of governance, power and knowledge (political economy approaches); the history of accountancy; and also ethnography or anthropology of the practices and culture of accountancy (Hopwood and Miller 1994; MacKenzie 2006). Second, we turn to the notion of governmentality to explore the relationship between discourse and practice or "techniques" in effecting power and authority. Third, the policy network concept of "epistemic community" is used to examine the nature of accounting expertise and its application to policy change. These reviews are necessarily brief: it is not the authors' intention to provide a full summary, but rather to consider how they might lend insight to the work of financial accountants in relation to climate change. For this reason we limit our attention to ideas from these literatures about calculation, measurement and expertise: who defines the problem and its solutions (a process necessarily involving forms of measurement and calculation), and how they generate the authority and capability to do so. We note at the outset that these literatures have different framings and conceptions of what is most important to study, eg for governmentality it is the day-to-day practices and techniques of government, whereas the notion of epistemic communities is more concerned with how particular groups of experts bring about change. However, we view these different perspectives as complementary, rather than conflicting.

Accountancy and Society
Scholars examining the relationship between accountancy and society aim to extend beyond narrow conceptions of accountancy, arguing that accountancy is not only relevant within the boundaries of a particular firm, but plays a constitutive role in social processes more generally. It is a broad literature, and what is of interest to us here is these close connections that have been demonstrated between accountancy and social processes (see Hopwood and Miller 1994), suggesting with regard to climate change what is pertinent is not just analysis of how the accountancy profession might be responding to increasing societal concerns about climate change, but also the role accountants might be playing in influencing how the problem is made sense of and dealt with. As Miller (1994:9) suggests, "Accounting could not and should not be studied as an organizational practice in isolation from the wider social and institutional context in which it operate[s]." In other words, social processes shape and are shaped by accountancy.

    This strand of critical accountancy and society research emerged during the late 1970s and early 1980s (with, for example, the foundation of the journal *Accounting, Organisations and Society* in 1976). Key research themes include: ethnography of accounting practices; political economy of accounting (eg looking at how power is exercised, at conflicting political and economic interests); and organisational design and environments (eg the notion of rationality which is pervasive in accountancy, but in reality accounting practice is much closer to *bricolage*, to "organised anarchies") (Hopwood and Miller 1994; Miller 1994). Miller's work is of particular relevance, outlining three ways of viewing accounting as a social and institutional practice: first, in seeing accounting as a *technology*—a way of intervening, giving visibility to events and processes, and of governing people; second, focusing on the complex language and meanings of accountancy—its *rationales*; and third, examining how things are made knowledgeable in economic terms through accountancy, the *calculative* aspects of accountancy (Miller 1994). It is striking how Miller's framework has parallels with governmentality approaches—discussed below—in suggesting accounting governance and practice can best be understood through examining discourse and technologies; it also has links to the concept of epistemic communities, through highlighting the role of expertise and knowledge.

To date this extensive body of work on accountancy and society, though relevant, has not been widely applied to the issue of accounting and climate change (for exceptions, see Cook 2009; Lohmann 2009; MacKenzie 2009). Such an approach would position carbon accountancy as intricately connected to wider societal debates about not just the environment, but also the relationship between markets and governments, the role of science and so on. Further, it would view carbon accountancy as having the capacity to shape society itself, as Miller explains:

> accounting is, above all, an attempt to intervene, to act upon individuals, entities and processes to transform them and to achieve specific ends. From such a perspective, *accounting is no longer to be regarded as a neutral device* that merely documents and reports "the facts" of economic activity. Accounting can now be seen as a set of practices that affects the type of world we live in, the type of social reality we inhabit, the way in which we understand choices . . . (Miller 1994:1, emphasis added)

Accountancy is meant in theory (according to professional codes of conduct) only to reflect "economic reality" and societal preferences and practices, but can in practice end up influencing them (Miller 1994; Miller and O'Leary 1994; Power 1994). The accounting and society literature is valuable therefore as a correction to the implicit assumption within the non-accountancy academic literature on climate change policy, politics and markets that accountancy is rule-based.

Accountancy and society perspectives might usefully highlight too the history of carbon accountancy, building on previous scholarship illustrating the path dependency and inertia in how accounting decisions are made, that is, once certain accounting practices are established they tend to remain (see for example Miller and O'Leary (1994) on the rise of standard cost accounting in the 1930s; Thompson (1994) on the emergence of double-entry accounting). With carbon accountancy still in its formative stages—with many critical decisions to be made—close attention to current governance processes and decision making is likely to have significant theoretical and policy impact. Further, scholars have drawn attention to the often subtle ways that power is expressed in decisions about detailed, technical accountancy rules (Miller 1994; Miller and O'Leary 1994; Thompson 1994). Accountancy can be a way of making things appear "anti-political" (after Barry 2005) and seemingly uncontroversial, but the technical debates about accountancy rules and standards sometimes involve intense power struggles. Because carbon accountancy rules (once decided) will potentially have a huge influence on company profits, liabilities etc, it is no surprise that it has been a site of conflict, a point returned to in conclusion.

We therefore now turn to review briefly two further bodies of literature that draw together these ideas about accounting practices more strongly and directly with the politics of policy change, including attention to issues of international politics, discourse, and the role of expertise.

Governmentality

A Foucauldian governmentality approach is another fruitful lens to consider issues of governance and authority in carbon accounting because of Foucault's longstanding interest in how power is expressed and can be identified through day-to-day practices and routine activities (including, for instance, calculation and book-keeping) (Foucault 1991, 2007). According to Foucault, since the mid-eighteenth century "government"—the self-regulation of behaviour, especially in the form of "self-control" by apparently freely choosing autonomous subjects—has been the main way states have sought to control their populations (Foucault 1991). Whilst the majority of work on governmentality has concentrated on the self-regulation of individual behaviour (amongst the general public), there is a growing interest in testing the applicability of such ideas to institutions such as non-governmental organisations and corporations (Hughes 2001; Sending and Neumann 2006); and it is this sub-set of the literature that we primarily draw upon here. The pervasiveness of financial accounting in modern society signals the potentially significant role it might play in "government". According to governmentality theory, scholars need to pay close attention: first, to how objects of government are defined and how problems are framed (termed "rationalities"), and second, how they are governed through "technologies" (Dean 1999). From a governmentality perspective, calculation and measurement are critical to governance processes, as Dean explains: "An analysis of government, then, is concerned with *the means of calculation*, both qualitative and quantitative" (1999:11, emphasis added).

For Murray Li (2007) in her governmentality analysis of development projects in Indonesia, the role of government is also primarily about calculation: it is concerned with making things (problems) into technical programmes that can be managed. Similarly to Dean, Murray Li identifies two key practices that are required to translate a government "rationale" or discourse into an explicit coherent policy programme: first, *problematization*—identifying the problem, the things that need to be rectified; and, second, *rendering technical*—a set of practices "concerned with representing 'the domain to be governed as an intelligible field with specifiable limits and particular characteristics . . . defining boundaries, rendering that within them visible, assembling information about that which is included and devising techniques to mobilize the forces and entities thus revealed.'" (Murray Li 2007:7; quoting Rose 1999:52). The two practices are, of course, intricately linked, for identification of a problem is linked to the availability of a solution (see also Kingdon 2003). A governmentality lens is especially relevant in thinking about carbon accounting because it brings to the fore the possibility that accounting technologies and practices can themselves influence wider discourse; it is a two-way relationship. As Murray Li (2007:6) explains, in the adoption of a governmental rationality:

> Calculation is central, because government requires that the "right manner" be defined, distinct "finalities" prioritized, and tactics finely tuned to achieve optimal results. Calculation requires, in turn, that the processes to be governed be characterized in technical terms. Only then can specific interventions be devised.

Through these ideas we begin to see how discourse, technologies and calculation are key to understanding the role of accountants and the accountancy profession in relation to climate change.

Epistemic Communities

The notion of an epistemic community was first elaborated upon by Haas (1992a, 1992b) and refers to a knowledge-based international community of experts, specifically a "network of professionals with recognized expertise and competence in a particular domain and an authoritative claim to policy-relevant knowledge within that domain or issue area," (Haas 1992b:3). The term was first used by John Gerald Ruggie in 1975, who coined it from Foucault's notion of an *episteme*, defined as "a dominant way of looking at social reality, a set of shared symbols and references, mutual expectations and a mutual predictability of intention" (Ruggie, 1975: 570, quoted in Verdun 1999). It was Haas who fully developed the concept, based on his observations of scientists working on the ozone hole and involved in developing the Montreal Protocol. According to Haas (1992b) four defining features of epistemic communities are: a shared set of normative and principled beliefs; shared causal beliefs; shared notions of validity; and a common policy enterprise. What can be usefully applied to the case of carbon accountancy is the idea of shared beliefs and values uniting a group of experts on a particular policy issue, which Haas and others (Gough and Shackley 2001; Litfin 1994) have argued stem from their professional culture and expertise. So an epistemic community perspective directs our attention to the professional culture and training of accountants, and the links between this culture and expertise and their beliefs about how to mitigate climate change, in turn reflected in detailed policy proposals. There is a notable contrast here with governmentality approaches which direct our attention more towards the day-to-day practices and techniques of government. Nevertheless, the two theories are complementary through their shared interest in expertise—whether that be associated with routine, widespread practices (governmentality) or collective pooling and application of knowledge (epistemic communities).

A political scientist, Haas positioned epistemic communities against a range of other policy network theories, which he argued gave too much attention to power in their explanations of (international) policy change, and not enough to the role of knowledge and expertise. Although there is a bias in the epistemic communities literature towards analysis of scientific communities, the definition of epistemic communities is sufficiently broad to encompass a range of types of professional expertise and knowledge, including accountancy. Indeed, there are some examples of the epistemic community concept being applied to financial policy making (see for example the paper by Verdun (1999) on European Monetary Union, which positions central bankers as an epistemic community). What is most important to our analysis here is Haas's ideas about the role of knowledge—how knowledge and expertise create personal connections—and how these in turn are applied to frame and solve particular policy problems: in this sense it applies well to accountants and the work they are doing in response to climate change, because of their highly specialised knowledge. Further, epistemic communities are seen as most important in conditions of uncertainty; and with climate change being a relatively new issue for accountants, and with its implications for the professions still unclear, the concept appears likely to have traction.

The progressive narrowing or "framing" of policy debates is a key function of an epistemic community, and there are links here with governmentality "rationales" and "problematization" (Dean 1999; Murray Li 2007). In his discussion of the ozone negotiations Haas sees the epistemic community (comprising mainly atmospheric scientists) as playing a vital role in setting the overall terms of the policy debate, drawing on its shared knowledge and expertise to identify and delineate the ozone hole problem and its solutions, as he explains:

> In the face of foreign policy decision makers' uncertainty about the causes of the problem [ozone pollution] and the possible consequences of action, the epistemic community was largely responsible for identifying and calling attention to the existence of a threat to the stratospheric ozone layer *and for selecting policy choices for its protection* (Haas 1992a:188, emphasis added).

More specifically Alder and Haas (1992) identify four mechanisms by which epistemic communities exert influence: first through *policy innovation*—in the initial framing of the issue; second, *policy diffusion*—whereby epistemic community members communicate ideas through their international contacts, by word of mouth and reports; third, *policy selection*, when policy makers seek out particular epistemic communities for policy ideas and support; and fourth *policy persistence*—the durability of ideas, beliefs and goals over time, which boosts an epistemic community's authority and credibility. With reference to the case under discussion—accountants and climate change—it appears that, at this relatively early stage of engagement, we are only witnessing the first two mechanisms: policy innovation and, to a lesser extent, policy diffusion. As we demonstrate in the main empirical section below, accountants have to date not been at the forefront of climate change action, but have nevertheless significantly increased their activities and interest in the problem in recent years and a number of international networks have emerged (eg the Climate Disclosure Standards Board, Accounting for Sustainability). Moreover, accountants have framed climate change in a way that makes their own expertise and knowledge (in calculation, measurement etc) highly relevant to the policy solutions, and in so doing have contributed to a wider framing of the issue as a matter of reshaping and extending market processes and existing corporate reporting procedures (and not radically altering or disrupting those processes and procedures).

In summary, in this necessarily brief review of the literature we have drawn together a diversity of concepts and theories which nevertheless have common themes with relevance for understanding carbon accountancy. First, we note that seemingly banal day-to-day practices and techniques can be central to processes of policy innovation and change, and highlight the ability of these practices (such as double-entry financial accounting of assets and liabilities) to influence the framing of policy debates, typically quite narrowly framed. Second, the literatures emphasise the importance of professional expertise and knowledge in developing policy responses, especially in conditions of uncertainty. Third, our brief review has drawn attention to the unusual current visibility of carbon accounting practices and techniques, because they are still being actively debated as climate change remains a relatively new policy problem—thereby demonstrating the value of this particular case study, to which we now turn.

Accountants and Climate Change

In this section of the paper we examine in detail precisely how accountants are engaging with climate change. We first provide a brief early history of carbon accounting (in the period from the late 1990s to 2005, termed "Stage One"), exploring how climate change first became a practical issue for financial accountants with the advent of the EU ETS, and demonstrating how at the time these financial carbon accounting activities bore little connection with wider (prevalent) debates about valuing the environment. Second, we assess the recent engagement of accountants with climate change ("Stage Two"—2005 to present). We examine the positioning of accounting as the natural professional home of carbon management, both in terms of how climate change has been defined by accountants, and the accounting technologies and techniques used to do this (how climate change has been "rendered technical" by the profession).

Our analysis concentrates on the role of financial accounting professional organisations, with a particular focus on those that have taken the lead in engaging with climate change. We have reviewed a range of relevant reports and internet sites, and explored the issues in depth through interviews (20 in total) with key industry players active in carbon accounting (particularly relating to the EU ETS; and the relaunch of the IASB/FASB Emissions Trading project).

Stage One: Late 1990s–2005 "Reluctant Engagement"

It was during the period around the turn of the century that climate change first became a technical issue for accountants. In the run up to the start of the EU ETS in 2005 a vigorous debate took place—amongst accounting specialists and largely "behind the scenes"—about how to incorporate EU ETS carbon credits (termed "EUAs") within financial accounts. This debate has been explored elsewhere (see Bebbington and Larrinaga-Gonzalez 2008; Cook 2009; MacKenzie 2009) and will not be rehearsed in detail here, but it is significant for our analysis because it was the first notable practical engagement of financial accountants with climate change (albeit it was restricted to Europe, and only of primary relevance to the companies active in the EU ETS, which covers 12,000 industrial installations in Europe), and second, because of how controversial and conflictual the process of reaching an agreement was (and continues to be—still, several years later, there is no official guidance on how to account financially for carbon allowances or credits).

In the run-up to the advent of the EU ETS, accounting guidance was issued by the International Accounting Standards Board via its International Financial Reporting Interpretations Committee (IFRIC): *IFRIC Interpretation 3: Emission Rights* (known as IFRIC-3) was published in December 2004. Whilst the detailed accounting recommendations of IFRIC-3 need not be recapitulated here,[4] what is notable is the amount of the controversy the recommendations generated. IFRIC-3 was eventually withdrawn because of negative reaction amongst major EU ETS participants (utilities, large industry emitters) on a number of grounds, including about where to account for carbon (with IFRIC-3 recommending some gains and losses to be reported in the income statement and some in equity, ie a "mixed presentation model"), and how to balance assets and liabilities (with IFRIC-3 recommending some carbon credits to be measured at cost and others at fair value, that is, a "mixed measurement model"; see Cook 2008; MacKenzie 2008). Since the withdrawal of IFRIC-3 there has been no international guidance on how to account for EU ETS credits, and a diversity of practices has emerged (Cook 2008; MacKenzie 2008; McGready 2008; PriceWaterhouse Coopers and IETA 2007). Since 2008, however, accountancy standard setters have become active again through the joint IASB-FASB Emissions Trading project (see below), which aims to resolve the situation through issuing clear guidance.

The IFRIC-3 withdrawal illustrates, we suggest, how accountants at the time viewed their role in mitigating the problem as largely technical and non-strategic. This finding rather calls into question the assumption amongst governmentality scholars that new discourse and practices necessarily have a particular agenda, with knowing actors driving it. The debate about IFRIC-3 took place behind closed doors— wider input was not canvassed, and in the debate few links were drawn to the more fundamental long-term implications of climate change for the accountancy profession (Deloitte 2009; PriceWaterhouse Coopers and IETA 2007). Indeed, this is despite concurrent wider debates on the principle of valuing the environment and accounting for environmental assets and liabilities, green reporting and corporate social responsibility

(Deegan and Blomquist 2006; Herbohn 2005; Villiers and van Staden 2006). During the period in question these two sets of debates—the wider societal debate about valuing the environment, and the detailed, technical debate about financial accounting and the EU ETS—did not intersect.

It is important, however, not to overstate the lack of engagement of accountants at the time with sustainability issues, including climate change. For example, the Institute of Chartered Accountants in England and Wales (ICAEW) in 2004 published a detailed report entitled *Sustainability: The Role of Accountants*, as part of its "Information for better markets campaign". Whilst not focused specifically on climate change, the report nonetheless engaged directly with the problem in a number of ways, particularly in positioning accountants (and their skills) as being pivotal in its management and identifying climate change as one of a number of "mega risks" that deserves attention (ICAEW 2004:18). The report, for instance, concluded in a chapter devoted to the issue of tradable permits:

> At present, very few professional accountants are familiar with the [tradeable permit] schemes . . . and there is a challenging opportunity for the profession to contribute to the development and implementation of policy at all levels, as well as standards for accounting and reporting . . . (ICAEW 2004:66).

The ICAEW 2004 report did not, however, at the time of its publication generate much response from the profession, as the manager with responsibility for sustainability issues at ICAEW explained in interview:

> I suppose what we were doing with [the 2004 report] was carving out a role for the profession, trying to identify it . . . and saying to members "Look, here is a role for you, and *tell us what skills we need to build for you so you can occupy it*"
>
> Interviewer: And what sort of a reaction did you get?
>
> Well, I'd say four and a half years ago the reaction was puzzled bemusement! I think members struggled—and still do to an extent—to see what their role is . . .
> (interview, October 2009).

The interviewee clearly identifies here an absence of skills and techniques through which accountants can respond to climate change, indicating thereby that during this "Stage One" of reluctant engagement the dominant process was of "problematization"—discursive framing of the problem of climate change—with the second key stage of "rendering technical" (the "how to do it") yet to occur. Results from ICAEW's 2003 survey of over 100 accounting firms on Social and Environmental Issues reinforce the interviewee's comments, with 63% of respondents at the time viewing environmental issues as largely irrelevant to the majority of their clients, and 47% judging environmental issues as outside the accountant's realm (ICAEW 2003:2).

The Canadian Institute of Chartered Accountants (CICA) was also active on environmental and climate change issues during the late 1990s and turn of the century. As early as 1993 CICA published a report on *Environmental Costs and Liabilities: Accounting and Financial Reporting Issues.* In 1997 it was a founding member of the Global Reporting Initiative (GRI) Steering Committee, and was heavily involved in drafting subsequent GRI Reporting Guidelines. Specifically on climate change accounting issues CICA took the lead in 2002 along with the International Emissions Trading Association (IETA) in sponsoring meetings of key industry and accounting players (the UK accounting standards board, the American Institute of Certified Public Accountants and FASB Emerging Issues Task Force and others) to consider greenhouse gas accounting in the run-up to the EU ETS (Casamento 2005). The detailed activities of this network of organisations, including private accounting firms and accountancy standard setters as well as accounting professional bodies, has been well documented elsewhere (Bebbington and Larrinaga-Gonzalez 2008; Cook 2009; MacKenzie 2009). We note here that this specialist, select group of accounting professionals working on the financial accounting issues associated with greenhouse gas emission reduction units bears some similarities with the concept of an epistemic community: a group of technical experts working to address a particular policy problem. Nevertheless, their climate change discussions did not have wider ramifications across the profession; their recommendations were not widely circulated or acknowledged, and in this sense the application of the epistemic community concept is limited. There was an initial framing of the issue by this group (the "policy innovation" first mechanism in Alder and Haas's proposal for how epistemic communities exert influence), but it appears to have stalled at the second mechanism, that of policy diffusion.

More broadly the response of accountants to the issue of climate change during the late 1990s and early twenty-first century fits with ideas from accountancy and society literature about the close links between accountancy and wider society. Arguably accountants and the accountancy profession were reflecting a lack of sustained, fundamental societal engagement with climate change at this time. There was consequently no proactive attempt within the accountancy profession to identify with the problem of climate change and "render it technical". Indeed, on the question of how precisely to engage with climate change in Stage One (late 1990s to 2005) the profession remained largely silent, even if a discourse about climate change was starting to emerge. Professional bodies should not be seen as always strategic: inertia, "tactics", and *bricolage* are common: these short-term, piecemeal professional responses to the problem of climate change were by no means atypical. In the section below, however, we turn to consider in detail how more recently climate change has been "made knowledgeable" in a more comprehensive and serious way by accountants.

Stage Two: 2005+ "Strategic Engagement"
Since around 2005 there has been a notable shift in the depth and pace of response of accountants to the problem of climate change. As well as the publication of a number of new climate change reports, newsletters and other initiatives by accountancy professional bodies, in 2008 the main global accountancy standard setter—the International Accounting Standards Board (IASB) relaunched its Emissions Trading Schemes Project, this time in conjunction with the US Financial Accounting Standards Board (FASB) (IASB 2008), in an attempt to resolve the longstanding ambiguity—since the withdrawal of IFRIC-3—about how to account for carbon credits. This new joint IASB–FASB project has a somewhat broader remit too: it is not just about the EU ETS but aims to address the accounting of all tradable emissions rights and obligations arising under any emission trading schemes—including New Zealand, Australia, and existing and proposed schemes in the United States—thus reflecting the international growth of emissions trading since the turn of the century.

In this section we consider in more detail precisely what work is being done by accountants—as experts, through discourse and techniques—to position themselves as managers of climate change, and to make carbon understandable to the profession. We explore how accountants have continued a process of defining and framing climate change, and the work the professional accounting bodies in particular are doing in positioning themselves as pivotal to delivering solutions to the problem. We then turn to consider the crucial next step in governmentality accounts of change: development of "technologies", the techniques and practices promoted by accountants as suitable for managing climate change. It is this process of "rendering technical" climate change that characterises Stage Two, and denotes a shift in the depth of seriousness of climate change to the profession.

Table 1 summarises a range of climate change initiatives (reports, programmes, activities) undertaken by international professional accounting bodies since 2005. As discussed, the theme of the role of professional accounting organisations' growing engagement in climate change emerged out of research interviews originally more directly

Table 1: Summary of international accountancy and auditing professional organisations' climate change activities, 2005+

Name of organisation	Membership details, remit and geographical coverage	Examples of climate change activities—2005+ (in date order)
ACCA—Association of Chartered Certified Accountants	Nearly 500,000 members and students globally, with an international network of 82 offices and centres. Originally formed in 1904 (as the London Association of Accountants), became known as ACCA in 1996. ACCA members are known as Chartered Certified Accountants, and are employed in industry, financial services, the public sector, or in public practice.	ACCA policy paper—"Going concern? A sustainability agenda for action". August 2008. Section 3 devoted to climate change, entitled "Climate change: how the accounting profession should respond" http://www.accaglobal.com/pdfs/technical/tech-gc-001.pdf Accounting and Climate Change Quarterly newsletter (April 2009+) provides information on: ACCA's own climate change work, media coverage; significant policy developments; conferences and training events etc. http://www.accaglobal.com/general/activities/subjects/climate/newsletter/ The Carbon Jigsaw (2009+)—a web-based initiative to provide ACCA members with appropriate tools and information on climate change. Several short briefing papers published in 2009 on topics such as: Carbon measurement, reporting and assurance (KPMG), and Carbon law (Baker and Mackenzie) http://www.accaglobal.com/general/activities/subjects/climate/projects/carbon COP-15 position paper (August 2009) statement outlining recommended action at the international climate change conference in Copenhagen. http://www.accaglobal.com/pubs/about/public_affairs/unit/global_briefings/cop15_aug09.pdf Carbon Accountancy Futures (forthcoming 2010). The themes (currently under development) are: Access to finance, Carbon accounting, Futures for audit and Narrative reporting. ACCA will fund research and encourage external partnering to do work under these themes. http://www.accaglobal.com/general/activities/subjects/climate/projects/accountancy-future

(Continued)

Table 1: Continued

Name of organisation	Membership details, remit and geographical coverage	Examples of climate change activities—2005+ (in date order)
ICAEW—Institute of Chartered Accountants in England and Wales	ICAEW is the largest professional accountancy body in Europe with 132,000 members and 9000 students. Over 15,000 members live and work outside the UK and its activities span 165 countries. ICAEW was incorporated by Royal Charter in May 1880. Members of the Institute are entitled to the description "chartered accountant" and to the designatory letters ACA or FCA.	*Sustainable Business Thought Leadership Programme* (2008+) aims "...to explore the importance of information in decision making", with a strong emphasis on climate change. *Business Sustainability e-learning programme* (2009) mainly focused on Corporate Social Responsibility (CSR), but with some climate change content. *Environmental Issues and Annual Financial Reporting* (2009) joint report with the UK Environment Agency; wide-ranging and comprehensive with particular attention given to the requirements of EU Directives, and with detailed advice and guidance for companies, financial report users, and auditors.
CIMA—Chartered Institute of Management Accountants	CIMA is the world's largest professional body of management accountants (management accounting combines finance and management with more general business techniques). CIMA is mainly UK based (with head office in London), but also has offices in Australia, China, Hong Kong, India, Ireland, Malaysia, Pakistan, Singapore, South Africa, and Sri Lanka. CIMA was founded in 1919 and has approximately 70,000 members.	Report *Emissions Trading and the Management Accountant* (2006) http://www1.cimaglobal.com/cps/rde/xbcr/SID-0AE7C4D1-81372D1C/live/tech_resrep_emissions_trading_and_the_management_accountant_2006.pdf *CIMA Position paper* "*Climate change calls for strategic change*" (March 2008) http://www2.cimaglobal.com/cps/rde/xbcr/SID-0A82C289-1DBAF8FB/live/tech_dispap_Sustainability_climate_change_Mar_2008.pdf *Position paper* "*All change*" (June 2008) *Editorial* examining role of management accountants in strategic management of climate change, published in CIMA's *Excellence in Leadership* report (2008) http://www.excellence-leadership.com/editorial/june08/CIM006-allchange.pdf

CICA—Canadian Institute of Chartered Accountants	The CICA has a link with the Bermuda Institutes/Order of Chartered Accountants, and together they have approximately 75,000 members and 12,000 students. The CICA has been active on sustainability issues for the last 20 years, and was a founding member of the Global Reporting Initiative.	Report *Building a Better MD&A (Management Discussion & Analysis): Climate Change Disclosures* (2008). A report with detailed advice to companies about what sort of climate change information companies should be reporting on, with close attention to investor requirements. It builds on a general document CICA (2004) *Management's Discussion and Analysis*, and was published in response to increasing demand for information about climate change (see Preface). http://www.cica.ca/research-and-guidance/mda-and-business-reporting/mda-publications/item12846.pdf Report *Climate Change Briefing: Questions for Directors to Ask* (2009). An in-depth report outlining the possible implications of climate change for businesses, and encouraging company directors to think strategically about their response to the problem. http://www.rmgb.ca/abstracts-directors-series/item28951.pdf
IAASB—International Auditing and Assurance Standards Board	The IAASB is a standard-setting body designated by, and operating independently under the auspices of, the International Federation of Accountants (IFAC). International Standards on Auditing (ISAs) are used by over 100 countries worldwide.	Consultation paper—*Assurance on a Greenhouse Gas Statement* (October 2009). Detailed consultation paper published with IFAC to seek views from practitioners and other stakeholders in order to develop an International Standard on Assurance Engagements (ISAE) for greenhouse gases. The IAASB intends to use the feedback from the consultation to develop an Exposure Draft of a proposed new assurance standard on GHG statements for release in 2010 http://www.ifac.org

focused on the financial accounting of carbon. This emergent finding was further investigated through desk-based discourse analysis of key reports, targeted interviews with accounting professional organisations, plus participant observation by Lovell in her role as a member of the Climate Disclosure Standards Board (CDSB) Technical Working Group.

As the data in Table 1 demonstrate, over the last few years there has emerged considerable interest in climate change amongst international accounting and auditing professional bodies. A range of reports have been published, such as the recent Association of Chartered Certified Accountants (ACCA) position paper on the 2009 United Nations negotiations on climate change and the International Auditing and Assurance Standards Board Consultation Paper (October 2009). In these reports and briefings climate change has been framed by accountants quite narrowly as a corporate problem—as something that the business community is responding to and will increasingly need to respond to—with the help of accountants, as the outline to ACCA's recent online "Carbon Jigsaw" initiative explains:

> Increasingly, ACCA members will need to understand what is happening globally in order to report emissions, monitor reductions or increases, and purchase or sell carbon offsets under emerging trading regimes. ACCA members will need to understand how the carbon crisis will affect businesses, and whether there are investment opportunities to exploit (ACCA 2009).

Notably, there are seen to be some positive aspects to dealing with climate change, such as the "investment opportunities" mentioned above. In this framing of climate change it is mainly private actors, including accountants, who are positioned as influential in initiating and managing change, as an extract from ACCA's material introducing the Carbon Jigsaw again clearly demonstrates:

> At some stage in the next 12 months ... every major business can expect to be asked about its greenhouse gas emissions and its mitigation strategy. Those asking such questions will be a variety of global and influential organisations such as the Carbon Disclosure Project, the Dow Jones Indices or FTSE Group, as well as governments and investors. *To respond to such questions and to demonstrate action, businesses will need to involve accountants. In the future, it will be the role of accountants to represent carbon-related actions in financial accounting terms in the annual reporting process* (ACCA 2009).

Note that the onus here is put on accountants to answer these questions, that is, to develop the methods and techniques to do so in the context of considerable ambiguity amongst the business community about how precisely to respond to climate change, a point returned to below.

Further, the profession has also sought to reassure others that accountants have successfully responded to similar types of problem in the past, and is therefore well equipped to deal with climate change:

> ... accountants are familiar with sustainability as a concept via a long history of dealing with capital maintenance. In wrestling with the concepts of income and capital, accountants have long been thinking in terms relevant to sustainability (ICAEW 2004:11);

and

> "The human race is at an important crossroad and will require all its famed ingenuity to continue to develop. Human history shows our ability to rise to challenges: think, for example of the programme of public health infrastructure in Victorian Britain, or the digital revolution of the last three decades. *The accounting profession must play its part in correcting the greatest and widest-ranging market failure ever seen* (ACCA 2008:8, emphasis added).

It is perhaps questionable whether the issues and problems listed here bear any real comparison with climate change (the digital revolution, for example) and this highlights again the rather positive spin put on climate change governance: the accountancy profession in the main conceives of climate change as something solvable, albeit with careful application of existing (accounting) skills, knowledge and expertise.

It is not just financial accountants but management accountants too who have begun to engage with climate change. In 2006 the Chartered Institute for Management Accountants (CIMA) published a hard-hitting report—*Emissions Trading and the Management Accountant*—framing the role of management accountants as vital in delivering climate change solutions:

> ... the European Union Emissions Trading Scheme (EU ETS) and the Kyoto Protocol are likely to demand new organisational competences to which management accountants will need to respond... *Management accountants will need to learn a new language associated with these initiatives if they are to be able to work alongside technical experts and to contribute to debates which are already affecting corporate agendas* (CIMA 2006:3, emphasis added).

CIMA's report is notable in acknowledging the "new language" of climate change that has emerged in the accountancy profession. The quotation above hints too at the objective of CIMA to stake a claim to the professional knowledge and expertise that accountants can bring to climate change, but also suggests a sense of professional unease:

that management accountants are not up to speed on the issue of climate change. Further, in noting the importance of working "alongside technical experts" the powerful position of financial accountants, with their skills and expertise to develop technical practices to respond to climate change, is emphasised. There has been a fast pace of change in recognition of the problem of climate change and its implications, and an evident desire on the part of the accountancy profession to "catch up". A number of accountancy professional bodies see an opportunity for accountants to take the lead in climate change, as demonstrated by these reports and via interviews. For instance, as the Secretariat to an international network of accounting organisations working on climate change explained:

> With climate change related disclosure being such a new discipline, that hasn't really yet established its own body of professionals, there's a rather fragmented approach within organisations where, you know, does it belong to the procurement department, the premises department, CSR [Corporate Social Responsibility]? It doesn't belong anywhere (interview, October 2009).

Thereby the Secretariat identified a "gap" in the professional governance (or "ownership") of climate change, which accountants are viewed as the most appropriate or "logical" profession to fill. These sentiments are echoed by the director of a charitable organisation set up to encourage accountants to be more involved in sustainability issues (including climate change), as she explains the reasons behind its establishment in 2006:

> Certainly the overriding aim in creating [the organisation] was twofold. Firstly recognition for sustainable development to be realised. Organisations really needed to be embedding it into their DNA, so I guess the values driven side of things. But a part of that is *what kinds of tools and processes are needed to really support that embedding process*. And just thinking from the accounting community perspective, what was their role? How they could play a part in creating the right kind of systems that would really support sustainable outcomes? (interview, October 2009).

Again, the references to identifying and developing appropriate "tools" and "processes" are striking: climate change has been successfully framed as a problem by the profession, now the focus is on developing new and modifying existing accounting practices in order to respond to it.

There are a number of international accounting and climate change networks emerging which might be viewed as epistemic communities, with members connected through their accounting expertise and shared professional culture and values. One such example is the Climate Disclosure Standards Board (CDSB), formed in 2007, whose Technical Working Group comprises accountants and representatives from the major international accountancy professional bodies (ACCA, ICAEW, CICA etc). The CDSB has the objective of developing a global framework for corporate reporting on climate change, and is pressing for climate change reporting to be integrated into mainstream financial reports (CDSB 2009). Another is the Prince's Accounting for Sustainability Forum, an international network of accounting organisations, albeit with a broader sustainability agenda (A4S 2009). Additional new coalitions and alliances may yet emerge, as the accountancy profession seeks external links with other professions and disciplines. The situation remains in a state of flux.

As noted, critical to the acceptability and uptake of a particular discursive framing of a problem are the techniques, practices and "technologies" that brought to bear on the problem. In other words, it is not just the discourse–society relationship that is important, but also the way accounting techniques and practices—modes of calculation—have shaped discourse, and profoundly influenced the accounting profession response. What distinguishes Stage Two from Stage One of the accounting profession's response to climate change is that in Stage Two a range of accounting "technologies" have begun to be associated with the prevalent accounting definition of climate change, and, as governmentality and epistemic community scholars would both argue, the response of the accountancy profession to climate change can only be fully understood by looking at discourse and technologies in tandem.

Accountants are seen as being in an influential position of authority within organisations, but also as having a very specific set of skills—numeracy, attention to detail, logic, understanding of assets and liabilities, corporate risks etc—through which to manage climate change (in keeping with the narrow definition of the problem). Interestingly, in the main responding to climate change is seen to involve the application of existing accountancy skills, rather than requiring the development of new ones, as illustrated in a CIMA Editorial on climate change:

Although significant change is needed [in response to climate change], these actions should ideally build on current business activities and skill sets. It is a matter of broadening the organisational mind set to encompass climate change issues, *applying existing skills and frameworks to address new challenges* (Doody 2008:9, emphasis added).

There is also a significant amount of ongoing activity and discussion about how to modify existing accounting software (which is critical to the day-to-day production of financial accounts—see Hatherly, Leung and MacKenzie 2008) in order to incorporate relevant climate change information, as one interviewee explained:

> There is such a huge stakeholder group with interest here that, whatever information is put into the public domain, it has to be useable. This then raises other questions on standardisation of how do you make that information useable? What sort of technology is required? Again, AICPA [American Institute of Chartered Public Accountants] has written quite extensively on use of XBRL.[5] And I know GRI [the Global Reporting Initiative] is looking at it. *Not only do you have to have the standards, you have to have the output, the presentation to translate that information into something that is actually useable* (interview, Secretariat to an international network of accounting organisations working on climate change, October 2009).

Accounting professional organisations view themselves as key players in identifying and modifying existing accounting technologies and practices in response to climate change, with the objective of making climate change understandable and relevant to their members.

Summary and Conclusions

In this paper we have analysed a number of professional accountancy organisation reports and drawn on interviews to illustrate how climate change has been framed as a particular type of problem by the accounting profession, with strong relevance for their skills and expertise. To date the dominant framing of the problem (and solutions) has been narrow: climate change is seen as a corporate problem, which is solvable with careful application of existing accounting approaches and techniques. We have shown how the accounting profession was initially rather slow to respond to the problem of climate change, with no significant engagement until the mid 2000s (albeit with a few notable exceptions— eg CICA and ICAEW). But the profession is now attempting, through the work of a number of accounting professional bodies, to rectify the situation and "catch up", positioning accountants as relevant, indeed crucial, actors in governing carbon.

We have explored how the problem of climate change has been moulded to fit within existing accounting discourse and practices, and in this sense climate change is not a distinctive problem: the new carbon economy represents "business as usual" for accountants. Further, we question to what extent the early actions taken by accounting organisations were strategic and deliberate, in the sense highlighted by governmentality perspectives, where discourse and practices are viewed as having a particular agenda, with knowing actors driving them. We suggest what occurred in Stage One around the turn of the century maybe more akin to "bricolage"—a more muddled and haphazard process. The significance of Stage Two (2005+), therefore, is the beginning of a more strategic engagement of the accounting profession with climate change, with signs of increasingly deliberate and careful positioning of accountancy skills and techniques as relevant. There is as of yet no clear epistemic community of accountants working on climate change, with the possible exception of the early group of experts working on emissions trading financial accounting issues (Stage One), but whose wider policy influence was limited (Casamento 2005; MacKenzie 2009). But there are signs that new accounting coalitions and alliances not solely focused on financial accounting may now be emerging, akin to prototype epistemic communities. For example, the Climate Disclosure Standards Board—with its Technical Working Group mostly comprising accountants—has developed an international voluntary carbon reporting standard, and further unexpected coalitions and alliances may result, in what is a new policy "space" for accountants generated by climate change.

We now turn to consider how this case of accountancy professional organisations fits with wider debates about carbon markets. Much of the criticism of carbon markets is not about the idea of putting a monetary value on carbon per se, but about whether it is better to control greenhouse gas emissions through setting emission standards (and then allowing trading, so-called "cap and trade"), or by charging the appropriate pollution taxes (Hepburn 2006). Weitzmann in his much-cited 1974 article "Prices vs quantities" makes a compelling economic case for pollution taxes. But, as argued elsewhere (MacKenzie 2007), the emergence of the cap and trade EU ETS was the only possible political option at the time in Europe. The key issue for this paper is that any economic mechanism for mitigating climate change (whether it be cap and trade or a tax) needs to pass through the filter of accountancy, and carbon accountancy therefore deserves close attention, both in policy and academic spheres.

We stress too that it is important to differentiate between different types of carbon market, something that critiques of market-based solutions to climate change often fail to do (FoE 2009; Smith 2007). Accounting climate change debates emerged, as we have shown, from the initial engagement of accountants with the EU ETS, and it is the EU ETS that has continued to influence the accounting profession as they have become progressively more engaged. Despite recent upsets (VAT carousel fraud, the "recycling" of CERs), the EU ETS appears to be working relatively well. We would suggest, however, that there is a much more compelling case for radical reform and overhaul of the other major global carbon market, the United Nations Clean Development Mechanism (CDM), mostly because of the damaging (political and atmospheric) inclusion of the industrial gas HFC-23 in the CDM, leading to perverse outcomes (Warra 2007).

The IFRIC-3 launch and subsequent withdrawal (Stage One) highlights how there is likely to be conflict in these technical accounting discussions when corporations feel strongly about an issue, typically when it affects their profits. It is interesting that with the relaunch of the IASB–FASB Emission Trading Schemes project conflict seems less evident. Indeed, recent interviews with accountants at major EU ETS companies have suggested a readiness for clear guidance from the standard setters (along the lines of IFRIC-3) because of a strong desire to make carbon accounting easier (reducing choice, thereby eliminating the current necessity of following a range of different national, international and corporate guidelines), and so that companies can be fairly compared with their competitors, creating a level playing field.

This situation in financial carbon accounting, where conflict appears to have eased, points to the possible benefits of continuing voluntary measures in the carbon disclosure (non-financial) aspects of carbon accounting. Conflict has not been evident there yet: what the CDSB and their counterparts are proposing is voluntary rather than mandatory corporate reporting. The accounting standard setters (IASB and FASB) will perhaps do well to take heed from the case of IFRIC-3 as to the dangers of rushing in too soon with mandatory guidance: ideally a transition to mandatory rules and practices will come at a point when corporations feel ready to welcome this clarity.

Acknowledgements

The authors would like to thank the Nuffield Foundation for supporting this research through an Early Career New Development Fellowship (2008–2013) "Fungible Carbon" held jointly by Dr Lovell and Professor MacKenzie. We would also like to thank all those who kindly agreed to be interviewed as part of this research for their time and valuable insights.

Dr Lovell would also like to thank the Association of Certified Chartered Accountants (ACCA) and the International Emissions Trading Association (IETA) for jointly funding research into the accounting of emission allowances in the EU ETS, and her colleagues on this project—Professor Jan Bebbington, Dr Carlos Larringa, and Dr Thereza Aguiar—for all their valuable input. She would also like to thank members of the CDSB Technical Working Group, in particular the Secretariat Lois Guthrie.

Endnotes

[1] Coding of interview transcripts and documents has been done in an inductive way, following initial leads arising from the data, and refining these over time into themes. All transcripts and documents have been analysed using the qualitative software "Atlas", which facilitates this type of "bottom up" coding. Examples of codes include: "expert knowledge", "carbon managed by accountants", "carbon as difficult to classify", and "uncertainty and discretion".

[2] The research project is called *Fungible Carbon*, and is a Nuffield Foundation New Career Development Fellowship (2008–2013), held jointly by Dr Heather Lovell and Professor Donald MacKenzie, investigating the tensions in developing a carbon commodity.

[3] Conducted jointly with Professor Jan Bebbington (St Andrews University), Dr Carlos Larringa (University of Burgos), and Dr Thereza Aguiar (Heriot Watt University).

[4] IFRIC-3 recommended that assets (in this instance the EU ETS allowances, EUAs) should be treated independently to the liabilities (ie obligations under the EU ETS). Accordingly, "netting off" of carbon assets and liabilities was not permitted. More specifically, IFRIC-3 gave guidance to the effect that emission allowances are intangible assets (whether allocated for free by government or purchased) and therefore fall under International Accounting Standard (IAS) 38. Further, allowances that are allocated for less than fair value should be measured initially at their fair value (ie market price), and the difference between the amount paid and fair value should be classified as a government grant and therefore accounted for under IAS 20 (Government Grants and Disclosure of Government Assistance). This "grant" should initially be recognised as deferred income in the balance sheet, and then subsequently recognised as income over the compliance period. In terms of liabilities, it was judged that a liability should be recognised as emissions are made, and that it should be a provision, falling under IAS 37 (Provisions, Contingent Liabilities and Contingent Assets). The liability should be measured at fair value, that is, the best estimate of the expenditure required to settle the present obligation at the balance sheet date.

[5] XBRL (extensible Business Reporting Language) is a formal data standard designed for application to financial accounting.

References

A4S (2009) *The Prince's Accounting for Sustainability Forum.* http://www. accountingforsustainability.org/output/Page150.asp (last accessed 14 October 2009)

ACCA (2008) *Going Concern? A Sustainability Agenda for Action.* http://www. accaglobal.com/pdfs/technical/tech-gc-001.pdf (last accessed 14 November 2009)

ACCA (2009) *Carbon Jigsaw.* http://www.accaglobal.com/general/activities/subjects/ climate/projects/carbon (last accessed 14 November 2009)

Alder E and Haas P M (1992) Epistemic communities, world order, and the creation of a reflective research program. *International Organization* 46(1):367–390

Barry A (2005) The anti-political economy. In A Barry and D Slater (eds) *The Technological Economy* (pp 84–100). New York: Routledge

Bebbington J and Larrinaga-Gonzalez C (2008) Carbon trading: Accounting and reporting issues. *European Accounting Review* 17(4):697–717

Casamento R (2005) Accounting for and taxation of emission allowances and credits. In D Freestone and C Streck (eds) *Legal Aspects of Implementing the Kyoto Protocol Mechanisms: Making Kyoto Work* (pp 55–70). Oxford: Oxford University Press

CDSB (2009) *Promoting and Advancing Climate Change-Related Disclosure: Exposure Draft.* London: Climate Disclosure Standards Board (CDSB)

CIMA (2006) *Emissions Trading and the Management Accountant.* London: Chartered Institute for Management Accountants (CIMA)

Cook A (2009) Emission rights: From costless activity to market operations. *Accounting, Organizations and Society* 34(3–4):456–468

Dean M (1999) *Governmentality: Power and Rule in Modern Society.* London: Sage

Deegan C and Blomquist C (2006) Stakeholder influence on corporate reporting. *Accounting, Organizations and Society* 31:343–372

DEFRA (2007) *Synthesis of Climate Change Policy Appraisals*. http://www.defra.gov
    .uk/environment/climatechange/uk/ukccp/pdf/synthesisccpolicy-appraisals.pdf (last
    accessed 3 July 2007)
Deloitte (2009) *Summaries of Interpretations: Ifric 3 Emission Rights*. http://www.
    iasplus.com/interps/ifric003.htm#withdraw (last accessed 23 November 2009)
Doody H (2008) All change. In CIMA (eds) *Excellence in Leadership* (pp 8–11). Bristol:
    Chartered Institute of Management Accountants (CIMA)
FoE (2009) *A Dangerous Obsession: The Evidence against Carbon Trading and
    the Real Solutions to Avoid a Climate Crunch*. http://www.foe.co.uk/resource/
    reports/dangerous_obsession.pdf (last accessed 10 December 2009)
Foucault M (1991) Governmentality. In G Burchell, C Gordon and P Miller (eds)
    The Foucault Effect: Studies in Governmentality (pp 87–104). London: Harvester
    Wheatsheaf
Foucault M (2007) *Security, Territory, Population*. Basingstoke: Palgrave Macmillian
Gough C and Shackley S (2001) The respectable politics of climate change: The
    epistemic communities and NGOs. *International Affairs* 77(2):329–346
Haas P (1992a) Banning chlorofluorocarbons: Epistemic community efforts to protect
    stratospheric ozone. *International Organization* 46(1):187–224
Haas P (1992b) Epistemic communities and international policy co-ordination.
    International Organisation 46(1):1–35
Hatherly D, Leung D and MacKenzie D (2008) The finitist accountant. In T Pinch and
    R Swedberg (eds) *Living in a Material World: Economic Sociology Meets Science
    and Technology Studies* (pp 131–160). Boston, MA: The MIT Press
Hepburn C (2006) Regulation by prices, quantities or both. *Oxford Review of Economic
    Policy* 22(2):226–247
Herbohn K (2005) A full cost environmental accounting experiment. *Accounting,
    Organizations and Society* 30:519–536
Hopwood A G and Miller P (eds) (1994) *Accounting as Social and Institutional Practice*.
    Cambridge: Cambridge University Press
Hughes A (2001) Global commodity networks, ethical trade and governmentality.
    Transactions of the Institute of British Geographers 26:390–406
IASB (2008) *International Accounting Standards Board (IASB)—Information
    for Observers: Emissions Trading Schemes; Board Meeting 20 May 2008*.
    http://www.iasb.org/NR/rdonlyres/92B01EDC-E519–431F-915F-0F33505D7DFD/
    0/ETS0805b03obs.pdf (last accessed 3 October 2008)
ICAEW (2003) *Environmental and Social Issues Survey: Research Report*. London:
    Institute for Chartered Accountants in England and Wales (ICAEW)
ICAEW (2004) *Sustainability: The Role of Accountants*. London: Institute of Chartered
    Accountants in England and Wales
IPCC (2007) *Climate Change 2007: The Physical Science Basis—Summary for
    Policymakers*. Paris: Intergovernmental Panel on Climate Change (IPCC)
Kingdon J W (2003) *Agendas, Alternatives and Public Policies*. New York: Harper
    Collins
Litfin K T (1994) *Ozone Discourses: Science and Politics in Global Environmental
    Cooperation*. New York: Columbia University Press
Lohmann L (2009) Toward a different debate in environmental accounting. *Accounting,
    Organizations and Society* 34(3):499–534
MacKenzie D (2006) *Working Paper—Producing Accounts: Finitism, Technology
    and Rule Following*. http://www.sps.ed.ac.uk/_data/assests/pdf_file/0010/3421/
    ProducingAccounts8Nov06.pdf (last accessed 13 September 2009)
MacKenzie D (2007) *The Political Economy of Carbon Trading*. http://www.
    lrb.co.uk/v29/n07/mack01_.html (last accessed 10 April 2007)

MacKenzie D (2009) Making things the same: Gases, emission rights and the politics of carbon markets. *Accounting, Organizations and Society* 34(3–4):440–455

McGready M (2008) Accounting for carbon. *Accountancy* July:84–85

Miller P (1994) Accounting as social and institutional practice. In A G Hopwood and P Miller (eds) *Accounting as Social and Institutional Practice* (pp 1–39). Cambridge: Cambridge University Press

Miller P and O'Leary T (1994) Governing the calculable person. In A G Hopwood and P Miller (eds) *Accounting as Social and Institutional Practice* (pp 98–115). Cambridge: Cambridge University Press

Murray Li T (2007) *The Will to Improve: Governmentality, Development, and the Practice of Politics*. London: Duke University Press

Power M (1994) The audit society. In A G Hopwood and P Miller (eds) *Accounting as Social and Institutional Practice* (pp 299–316). Cambridge: Cambridge University Press

PriceWaterhouse Coopers and IETA (2007) *Trouble-Entry Accounting—Revisited*. London: PriceWaterhouse Coopers (PWC) and International Emissions Trading Association (IETA)

Sending O J and Neumann I B (2006) Governance to governmentality: Analysing NGOs, states and power. *International Studies Quarterly* 50:651–672

Smith K (2007) *The Carbon Neutral Myth: Offset Indulgences for Your Climate Sins*. London: Carbon Trade Watch

The Aldersgate Group (2007) *Carbon Costs: Corporate Carbon Accounting and Reporting*. London: The Aldersgate Group

Thompson G (1994) Early double-entry bookkeeping and the rhetoric of accounting calculation. In A G Hopwood and P Miller (eds) *Accounting as Social and Institutional Practice* (pp 40–66). Cambridge: Cambridge University Press

Verdun A (1999) The role of the Delors Committee in the creation of the EMU. *Journal of European Public Policy* 6(2):308–328

Villiers C D and van Staden C J (2006) Can less environmental disclosure have a legitimising effect? Evidence from Africa. *Accounting, Organizations and Society* 31:763–781

Warra M (2007) Is the global carbon market working? *Nature* 445(8):595–596

Weitzmann M L (1974) Prices vs quantities. *The Review of Economic Studies* 41(4):477–491

Chapter 7
Realizing Carbon's Value: Discourse and Calculation in the Production of Carbon Forestry Offsets in Costa Rica

David M. Lansing

Introduction

In 2004, a group that included scientists, economists, indigenous leaders, and state bureaucrats began work on an agricultural development project among indigenous Bribri and Cabécar smallholders in southeast Costa Rica. This project's original goal was to revive cacao agroforestry practices by linking the production of agroforestry landscapes with an emerging global commodity—the carbon forestry offset. Specifically, project developers wanted to create a carbon forestry offset under the Clean Development Mechanism (CDM), where indigenous land users would receive a carbon payment for converting their pesticide-intensive plantain fields to more carbon-intensive cacao agroforestry systems. During the course of implementing this project, however, its goals shifted. After project managers completed cost–benefit calculations of various land use practices, they determined that the opportunity costs of switching from plantains to cacao agroforestry were too high for carbon financing to induce this type of change. Instead, their calculations revealed that carbon credits are better positioned to encourage the abandonment of swidden (slash-and-burn) systems of agriculture. In performing these calculations, project managers conceived of recently fallowed land, or *rastrojos*, as agricultural spaces that lack economic value, but which have high levels of carbon sequestration potential.

The New Carbon Economy, First Edition. Edited by Peter Newell, Max Boykoff and Emily Boyd.
© 2012 David M. Lansing. Book compilation © 2012 Editorial Board of Antipode and Blackwell Publishing Ltd.

Today, the project's largest single source of carbon storage now comes from allowing *rastrojos* to revert to secondary forest over a period of twenty years (CATIE 2006:80). In short, as a result of these calculations, the project's trajectory shifted from its original goal of *reviving* one form of indigenous agriculture to *replacing* another type.

While the stated goal of the carbon project was to promote ecologically friendly forms of land use in a culturally sensitive way (see Guzmán 2006), this project's calculations helped produce a final result that could potentially run counter to these aims. For example, swidden systems rely on a field rotation where recently harvested plots (*rastrojos*) are allowed a number of years to recover. This form of agriculture is used to grow maize, rice, and beans, subsistence staples that are commonly managed by women and which can provide an important hedge against price swings in basic foods (see Borge and Castillo 1997). Removing *rastrojo* plots from the long-term cycle of subsistence plantings would allow less time for currently utilized plots of maize and beans to recover before the next planting, potentially inducing long-term ecological damage and threatening the food and livelihood security of households. This example seems to follow the pattern found in other cases of commodifying nature, where the articulation between the abstract representations required of commodification and the socio-ecological complexity of locally produced natures result in projects that can produce negative social and environmental consequences in the long run (see Castree 2003; Robertson 2006; for carbon payments see Boyd 2009; Brown and Corbera 2003).

In this paper, I analyze the cost–benefit calculations of this project in order to examine why and how these "failed articulations" between the universal demands of capital and locally complex socio-natures occur. I do so by examining the relation between discourse and value in the production of a carbon offset commodity, where I treat the cost–benefit calculations of this project as discursive statements that enable the creation of value. Doing so, I make two central arguments. First, I argue that these calculations are situated within a wider discursive formation concerning indigenous bodies and their relation to agriculture, where indigenous agriculture can and should be improved in particular ways. This discursive formation allowed for the carbon offset project to emerge as a solution framed by a specific socio-spatial problematic, where the "problem" of indigenous agriculture is posed in ways that call forth the production of commodified agricultural spaces as the rational solution. I posit that carbon offsets emerged within a discursive formation where the "indigenous land manager" and "indigenous agriculture" are spoken about in ways where the two *should be* aligned in ways that maximize the economic efficiency of the former through the spatial optimization of the latter. This way of speaking about agriculture and indigenous peoples, in turn, gives rise to commodified spaces, such as carbon offsets, as the most desirable way to make such an alignment occur.

Second, I draw on Marxian value theory to understand why these calculations, as discursive statements, were able to open up some spaces (*rastrojos*) for commodified carbon while foreclosing on others (cacao agroforestry). I argue that due to the global climate regime within which this particular carbon offset is situated (ie the Kyoto Protocol), these calculations were necessary to establish an offset's use value. In this case, the usefulness of a spatially bound, carbon forestry offset ultimately derives from its ability to contribute to a globally coordinated management of the world's carbon cycle. This means that a carbon offset's use value is not found in its qualitative characteristics, but instead, is found in the quantitative representations of its spaces. Thus, the project's shift to *rastrojos* did not derive from calculations needed to make these spaces commensurable for *exchange*, but rather, the project's final form resulted from calculations that were needed for these spaces to achieve commensurability with a globally conceived carbon cycle. This is a discursive transformation that is necessary for offsets to become *useful as commodities* within the framework of the Kyoto Protocol. In short, I seek to explain this project's origins and trajectory through the relation between the local discursive formation that allows for these calculations about indigenous agriculture to be understood and taken seriously, and the global orientation of this commodity that required a process of valuation that ultimately altered the specific spaces that were available to be commodified.

My arguments in this paper are meant to help work through a problem that Noel Castree recently identified in geographic scholarship on neoliberalism and nature, where he points out the difficulty of drawing out generalizable principles from the specific case studies of the ongoing "neoliberalization" of the nonhuman world (see Castree 2008a:137–141). Castree himself tries to overcome this difficulty by relating this process to any number of "environmental fixes"—efforts by capital or the state to resolve fiscal, political, or accumulation contradictions through the implementation of market-based forms of environmental governance (Castree 2008a, 2008b). Doing so, Castree theorizes the proliferation of "neoliberal natures" as a global-scale phenomenon, where he seeks to answer the question: "Why are human interactions with the nonhuman world being 'neoliberalised' *across the globe?*" (Castree 2008a:131, italics mine). In contrast, I inquire into how a neoliberal project comes to be desirable at a *specific site*, and point to the universal logics of capitalist value in order to explain the final form that a project ultimately takes.

Doing so, I draw on a wealth of critical research from development studies that has identified the emergence of "development" as a discursive and material project whose contours are constituted, in part, by the process of capital accumulation (eg Ferguson 1990; Gidwani 2008; Wainwright 2008). This is an ongoing process that is productive of development subjects, where "development" becomes a naturalized, and desired, goal for the diverse subjects that fall within the purview of liberal capitalism (Gidwani 2008). This critical approach has been broadly taken up by a number of scholars toward understanding "green development" projects as well, where they have shown how subjectivities become linked with the environment to produce environmental subjects whose interests concerning natural resource management become closely aligned with those of the state (eg Agrawal 2005; Birkenholtz 2009). Under this process not only are the subjects of development produced, but their insertion within processes of capital accumulation and the governance goals of the state means that an ongoing supply of various objects and sites of development are continually produced.

In this paper, I take the commodification of carbon to be a discursive process of development in which specific sites and objects enter into a field of intelligibility in a manner that allows for some ways of understanding them while foreclosing on others. Specifically, this is a process by which value is produced in the spaces of indigenous agriculture through its discursive attachment to carbon. While I broadly agree with the idea that the commodification of carbon can be read as a type of environmental fix, as Castree might suggest (Castree 2008a; see also Bumpus and Liverman 2008), I resist the idea that these spaces have come to be desirable as commodified spaces *because of* the extension of global-scale capitalist processes to local sites. Instead, I ground my analysis in discursive formations of development, and show how nature's continued commodification is a process by which specific spaces, natures, and bodies come to be represented in ways that allow for neoliberal projects to emerge as the logical solution to longstanding development problems, with their final form ultimately conditioned by the requirements of capitalist value.

The paper proceeds as follows. In the next section I first explain the CDM's requirement of "additionality", and how the project's cost–benefit calculations help establish the additionality of a carbon offset. Then, I explore the discursive formation within which these calculations are situated by providing a history of development interventions in the Talamanca region, and the emergence of three discursive objects: the indigenous land manager, indigenous agriculture, and cacao agroforestry. I argue that the rules of formation around how these objects are spoken about have opened a specific socio-spatial problematic of development that carbon is positioned to solve. In the following section I analyze the cost–benefit calculations that underpinned this specific CDM carbon project in Talamanca, and how project managers came to treat *rastrojos* as atomized spaces of carbon-value potential. In the penultimate section I elaborate on what it means for additionality calculations to be discursive statements that enable value by placing these calculations within a Marxian understanding of value. Specifically, I draw on Kojin Karatani's (2003) interpretation of Marx in order to argue that these calculations can be understood as discursive statements intended to establish an offset's use value. I follow this argument with a discussion of why a carbon offset's position as a local space within a global regime of climate management ultimately led to the conflation of additionality calculations with this commodity's use value. I conclude by highlighting the analytic benefits to understanding the relation of discourse and value to carbon's commodification.

Development Discourse: The Problem of Agriculture and the Solution of Carbon

The cost–benefit calculations I described at the beginning of this paper were done in order to comply with the CDM requirement of "additionality", a term that encompasses a broad range of evaluative approaches that are designed to ensure that carbon financing will produce a project does not subsidize *status quo* forms of land use (Michaelowa 2005). These calculations were performed in order to show that the project is financially additional, which means that carbon finance is needed for a project to occur (Bumpus and Liverman 2008). To meet CDM approval, project managers needed to demonstrate that certain types of land use changes were not possible under current market conditions, but *would* be possible with the influx of carbon financing. Thus, project managers needed to compare the current rates of profitability of different forms of land use with their levels of potential carbon sequestration in order to show what potential types of land use "switching" would require carbon financing. In this way, the cost–benefit calculations served to quantify an imagined future where carbon finance could potentially change the future decision making of indigenous farmers.

Rather than evaluate these calculations in terms of their claims to truth, I instead broadly follow Foucault's archaeological method (Foucault 1972) and treat these calculations as statements that occur within a discursive formation, where speaking about indigenous agriculture in terms of quantified cost–benefit tradeoffs is an intelligible way of speaking to the diverse subjects that help bring a carbon offset into being: scientists, CDM and state bureaucrats, offset consumers, and indigenous leaders. In other words, I ask about the conditions that allow for such calculations to be taken seriously among these diverse subjects, and the effects that such a way of speaking can have. While additionality calculations are necessary to meet the requirements of the CDM, the calculations themselves occur within a discursive formation that is, in part, grounded in a local history of development interventions in the region. Thus, I analyze these calculations as discursive statements in order to understand how indigenous agriculture came to be considered as a site of carbon storage at all.

In the remainder of this section, I examine the rules of formation that define the emergence of three discursive objects that have consistently been at the center of development interventions in Talamanca since the 1980s: the indigenous land manager, indigenous agriculture, and cacao agroforestry. These objects have been of special concern since the *moniliasis* fungus (*Moniliophthora roreri*)—microbial spores that attach themselves to cacao pods and render them inedible—swept through the Talamanca region in 1979, which at the time, was the country's largest cacao-producing region (Dahlquist et al 2007). That was an event that, within a few years, transformed the Talamanca region from Costa Rica's largest producer of cacao—a crop that was grown using few chemical inputs and a diversity of shade trees—to a region that primarily produces chemical-intensive plantain monocultures (Polidoro et al 2008). This agro-ecological transformation marked the beginning of a period of state-led and internationally financed agricultural development projects aimed at reviving cacao agroforestry in the area that has continued until the present. This event also marked the emergence of a period of intensive inquiry into a particular conception of indigenous agriculture.

During this time, the "indigenous land manager" and "indigenous agriculture" became increasingly common objects of study. In general, development projects, as well as academic writings during this time, tended to be oriented around two key "problems" associated with indigenous agriculture: the abandonment of ecologically friendly, "traditional" forms of land use such as cacao and banana agroforestry (eg Somarriba and Beer 1999), and the looming specter of unsustainable population growth (eg Borge and Castillo 1997; Vargas Carranza 1985). These two problems were often linked. Writings during this period argued that indigenous agriculture is unable to keep up with the demands of a growing population without recourse to increasingly ecologically destructive, and modern, forms of land use (Borge and Castillo 1997; Borge and Laforge 1995; Castillo 1999). The proposed solutions to these problems were often centered on two things: the importation of new technologies—usually hybrid, monilia-resistant varieties of cacao (eg Beer 1991); and/or increasing the economic value of "traditional" crops. The latter to be done through either better marketing of "organic" products for export (eg Hinojosa Sardan 2002), or by intermixing more valuable plant species, such as lumber and spice trees, within the spaces of "traditional" agriculture, resulting in a updated form of agroforestry (see Beer 1991; Somarriba 1997).

Projects designed to improve indigenous agriculture were often linked to improving the indigenous body as well. This relation can be seen in the stated objectives of a large Dutch-financed development project in this area in the mid 1990s, Proyecto NAMASÖL, whose goal was to: ". . . try and create a process of technological change within the evolutionary context of the indigenous culture of Talamanca; which means triggering the potential transformation of the *Talamancan producer*" (Borge and Laforge 1996:3, italics mine). This project's overall goal of promoting "sustainable development" was wide ranging, and included strengthening the institutional capacity of the area for managing and protecting a national park (La Amistad National Park), introducing educational and health programs to the area, and introducing new technologies and best management practices for promoting sustainable agriculture (organic fertilizer and pesticide, monilia-resistant trees etc).[1] Below, I examine more closely the consultancy report for this project (see Borge and Laforge 1996) and how it conceptualizes the "Talamancan producer" and its relation to the "agricultural system". I do this to illustrate the discursive relationship between the indigenous body and agriculture and how the two are spoken about within the contemporary context of development interventions in this region.[2]

Co-authored by an agricultural economist and a cultural ecologist, this report advances the following hypothesis concerning agriculture in Talamanca: "there exist two systems of production in opposition:the traditional system and the outside system" (Borge and Laforge 1996:5).[3] The document ultimately concludes that, instead of existing in opposition, these two systems are complementary, and are linked through the rationality of the "indigenous producer". Figure 1 is redrawn from the report, and it illustrates how the two systems are tied together through the cultivation of corn. Under this conceptual scheme, the "indigenous producer" grows corn (a "traditional crop"), which allows him/her to throw work parties, where workers are "paid" by being given Chicha, a mildly alcoholic beverage brewed with fermented corn. Using an econometric analysis, the report demonstrates that such work parties are a less expensive way for the producer to access labor for their cash crops than paying daily wages. Thus, the economic efficiency of this form of communal labor means that traditional agriculture provides an economic subsidy for modern cash crops.

Why has such a delicate balance between traditional and modern halves of agriculture emerged? To answer this question the authors go through great lengths to describe a holistic picture of the indigenous producer, arguing that his or her goal is not merely to maximize

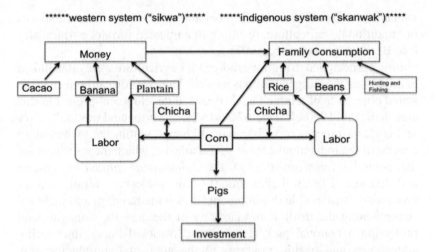

Figure 1: The indigenous agricultural system. Here, Corn (middle) acts as the hinge between the "western system" (left side) and the "indigenous system" (right side) where it allows households to brew Chicha, and throw work parties, which is deemed an economically efficient way to access labor for growing cash crops (source: figure redrawn from the NAMASÖL consultancy report; Borge and Laforge 1996:29)

profits but to improve one's "wellness", a definition that includes "non-economic" goals such as "good relations with neighbors", "good health", and "cultural expression" (see Borge and Laforge 1996:17). Presumably, this expansive understanding of wellness explains why such a diverse system, with its traditional and modern elements, exists today. Nevertheless, when it comes to explaining how the indigenous agricultural system is able to function, with its simultaneous orientation towards subsistence agriculture and cash crops, the indigenous body emerges in the form of the abstract rational economic actor. In this case, the indigenous land manager's economic rationality toward labor emerges as the critical hinge upon which a complex, and precarious, balance between the modern and traditional parts of agriculture is maintained.

Despite the perceived rationality of the indigenous body, development writings on Talamanca have consistently posited this system as a problem. The "modern" half of the system—plantain monocultures—is often referred to as an ecologically unsustainable system that "demands a lot from the soils" (Borge and LaForge 1996:37), and results in pesticide and fertilizer run-off (eg Polidoro et al 2008). Meanwhile, the "traditional" half is often argued to be insufficiently productive to meet projected population growth trends (Borge and Villalobos 1995; Castillo 1999). In other words, each part of the agricultural system is unsustainable in different ways; where the modern half of the system is unsustainable ecologically, the traditional half is characterized as unable to accommodate the needs of a changing society. While the indigenous producer is posited as economically rational, this rationality has nevertheless produced a two-part agricultural system that is inadequate to the twin problems of population growth and environmental decay. The report's conclusion follows the implication of nearly all development and extension literature of the last 30 years in this region when it calls for the "improvement" of the agricultural system. Since the arrival of the *monilia* fungus, "improvement" of the agricultural system has almost always coalesced around the revival of a third discursive object that has been critical in the framing of development interventions in this region: the cacao tree.

Cacao and Carbon

The cacao tree has been a critical object of intervention for the Tropical Agricultural Research and Education Center (Spanish acronym: CATIE), a regional agricultural development institution that was responsible for implementing the CDM carbon offset project in Talamanca and an institution with over 20 years of experience working with indigenous peoples in Talamanca. Beginning in the late 1970s, "agroforestry"—understood as the intermixing of crops and tree species—emerged worldwide as a specific object of promotion (Nair 1993; Schroeder 1999). CATIE joined this approach to agricultural development and began promoting agroforestry as a solution to the perceived problems of overpopulation and desertification in tropical areas (CATIE 1995). The perceived ability of agroforestry to solve these problems is captured by Eduardo Somarriba, who was the senior scientist in charge of the Talamanca carbon project. In 1981 he wrote the following in an introduction to a study on agroforestry for CATIE:

> The exponential growth of the population in tropical areas has led to increased demand for food and expansion of area under cultivation. This expansion has put pressure on tropical soils which will not support intensive use. At the same time the local and worldwide demand for forestry products increases and establishes a conflictive situation between land use options... An appropriate alternative would be agroforestry systems, which would fundamentally give the moist tropics a forestry vocation (Somarriba 1981:2).

In this passage, agroforestry is posited as a desirable method of rural development because it neatly solves a number of problems at once by being both spatially optimized and economically efficient. While these virtues of agroforestry were being promoted by scientists at CATIE, longstanding cacao agroforestry practices in Talamanca were quickly vanishing because of the spread of the *monilia* fungus.[4]

It is within this discursive and material context that CATIE initiated its first major agricultural undertaking in Talamanca. Its first project began in 1984 and was centered on introducing monilia-resistant, hybrid varieties of cacao trees (Dahlquist et al 2007). In 1987, a second project, a collaboration with the German government, was an effort to introduce more economically valuable forms of agroforestry practices, such as increasing the density of valuable shade trees in cacao plots and encouraging farmers to plant trees along their property lines (Beer 1991; Somarriba, Dominguez and Lucas 1994). CATIE's third major intervention into promoting cacao in Talamanca began in 2002 and marks a shift away from maximizing the economic efficiency of cacao agroforestry, and instead expanded to understanding and promoting the ecological value of agroforestry in the region. The "Cacao and Biodiversity" project, a World Bank financed project, sought to improve biodiversity conservation in cacao farms in this region (Somarriba et al 2004). While still trying to increase the economic and biological viability of cacao trees through many of the same methods as before (such as introducing *monilia*-resistant hybrid trees), this new project changed its mandate and linked the value of cacao agroforestry to wider ecological benefits in the region.

I provide this brief sketch of CATIE's history of cacao promotion in Talamanca, and the justifications of its work, in order to mark the general discursive rules that have guided how cacao agroforestry is spoken about. As a development tool, agroforestry's attractiveness lies in its dual features as an economically efficient use of space and an ecologically sustainable form of land use. Because of these two characteristics, cacao agroforestry is conceived as a solution to the problems of overpopulation and degradation. Later, cacao agroforestry became a geographically connective space as well; a farming system that promotes biodiversity and can provide an anthropogenic "link" to parks and wildlife corridors (eg CATIE 2006). Nevertheless, despite these advantages, "cacao agroforestry" is a system that requires development interventions if it is to flourish, because in Talamanca it remains biologically and economically unviable. Its susceptibility to *monilia* requires the introduction of resistant varieties of cacao, and its economic non-viability has been the impetus for efforts to make cacao agroforestry more profitable, either through increasing its exposure to organic markets (Hinojosa Sardan 2002) or by increasing the value of the spaces of a cacao plot itself, namely through increasing the intensification of commercially valuable trees (Beer 1991; Borge and Laforge 1996). In short, cacao agroforestry is posited as an ecologically friendly and efficient use of space, but with specific biological and economic constraints that require development interventions for it to spread in Talamanca.

It is within this development puzzle that the discursive objects of the "indigenous producer" and "indigenous agriculture" emerge as key pieces. Recall that while the indigenous agricultural system as a whole follows a particular cultural and economic logic, its reliance on "modern" crops like plantains renders it ecologically unsustainable (eg Harvey, Gonzalez and Somarriba 2006). Promoting cacao agroforestry can potentially solve this ecological problem. The indigenous producer's economic rationality, however, means that the success of a project requires an increase in cacao's relative value. According to development writings during this time, indigenous farmers, in their role as economically rational maximizers, tend to favor the most profitable crops (eg Hinojosa Sardan 2002). This means that maximizing the relative value of cacao agroforestry was a necessary step towards increasing its use (eg Somarriba 1993). In other words, before cacao agroforestry can solve the "problem" of indigenous agriculture its properties need to align more closely with the economic rationality of the indigenous body. CATIE's efforts to introduce commercially exploitable crops and to promote denser stands of timber trees were efforts to "improve" cacao agroforestry so just such an alignment could occur.

Since the 1980s, the "problem" of indigenous agriculture has been defined in terms of it being ecologically and economically unviable. This is a problem that two decades of agricultural development projects have tried to solve. I argue that both the framing of the problem and its solutions are emergent from a defining problematic, which is understood here as the system of reference points and relations that open some answers and foreclose on others (see Althusser 1979 [1965]). While debates on rural development in this area are complex, and often contradictory, the problematic of development in Talamanca can be read, in part, as the incongruence between these three discursive objects: the economically rational indigenous body, the ecologically unsustainable "agricultural system" and the ecologically friendly yet economically unviable cacao tree. And it is within this problematic that carbon offsets emerged as yet another way to align these three discursive objects. As we will see, once this effort at promoting cacao intersected with the process of creating value through carbon, different spaces altogether emerged as "ready" for development.

Enter Carbon Offsets: The Cost–Benefit Calculations

Like most agricultural development projects in the region, the original goal of this particular carbon offset project was to promote the use of cacao agroforestry. The original idea behind this project was that, with the arrival of CDM carbon offset financing, cacao agroforestry would become an economically viable space; and along with continued efforts to introduce monilia-resistant hybrid varieties of cacao, both biological and economic constraints to planting cacao would be overcome and farmers would be able to make the switch from growing monoculture plantains to planting cacao all while still following their rational economic interests (anonymous interview 2008; CATIE 2006; Segura 2005). In short, carbon payments would make cacao farming more profitable (thus overcoming the economic constraints) while continued efforts to give away resistant trees would overcome the biological constraints to cacao production.

In order to establish an offset, however, project developers needed to prove that just such a "switching" would take place because of the influx of carbon financing—the "financial additionality" requirement discussed above (Michaelowa 2005; Segura 2005). To help meet this requirement, project planners calculated the opportunity costs of switching from one form of land use to another. To do this, the labor, material, and transportation costs for each form of land use were estimated, along with typical production rates and market prices for each crop. These data were used to establish the net present value of each type of land use. In addition, each form of land use was assigned a carbon fixation rate based on biomass and soil measurements done by the project, which followed methodologies derived from the CDM. Finally, opportunity costs were calculated by comparing changes in the net value of land use after switching, divided by the net change in carbon fixation resulting from the land use switch. The resulting calculations showed the minimum carbon payment needed to induce a farmer to change from one form to the other. Under this reasoning a farmer would need an extremely large carbon payment (US$960/tonne) to cover the opportunity costs involved in a switch from plantain to cacao agroforestry (Segura 2005:34).

Just as different types of greenhouse gases must be made commensurable for carbon markets to function (eg MacKenzie 2009), project developers needed to be able to compare qualitatively different, yet often interconnected, forms of land use in terms of quantified cost–benefit tradeoffs. This meant discursively treating these spaces as separate, standalone forms of agriculture. While a typical *rastrojo* field was once a corn field, and will one day likely become a rice field, each one of these types of land use were considered atomistically separate, and "frozen" in time for the purpose of calculating their value-to-carbon ratio.

Project planners had very good reasons for doing this, since the carbon offsets were meant to induce future changes, so the carbon value of each form of land use had to be considered as they exist at the present moment. As one of the project planners explained:

> ... [indigenous farmers] have a fallow cycle and a crop cycle, but with this type of system they have a number of options, they have the option of rice, they have the option of maize, they have the option of beans, so we tried to analyze the incomes that each one of these choices generates ... in order to see the minimum that we would have to pay them if one producer is working with maize and if another is working with rice, that's why we did what we did, although obviously it's a complete cycle (anonymous interview 2007).

In other words, in order to understand how carbon financing would impact the future pathways of individual agricultural spaces, project planners had to bracket the long-term relationship of one space to another and consider the value of each space at the moment of an offset's creation. Figure 2 shows how these spaces were analytically "mapped" by project developers in relation to their value and carbon content, showing where each space falls in relation to the others. In order to show these relationships, each of the steps of the swidden cycle (crop–fallow (*rastrojo*)–secondary forest–crop) are separated so that they may

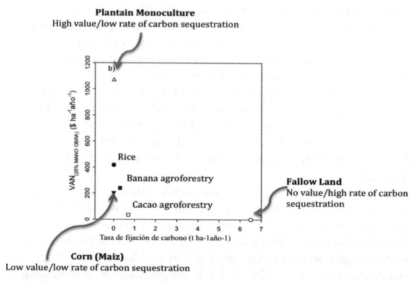

Figure 2: Annotated graph from the final project design document of the Talamanca Carbon Project. Net present values of each type of land use are plotted in relation to their rate of carbon sequestration (x-axis) and their net annual value per hectare per year (y-axis). Net annual value figures assume the household hires 25% of their labor (source: Segura 2005)

be related to all other forms of land use in the region in order to calculate the ability of carbon finance to induce "switching".

To make this graph, an idealized form of each type of land use was discursively severed from its ecological and economic links to other forms of land use so that these ideal types might be compared with each other in terms of their present economic value and future carbon potential. By discursively marking *rastrojos* as a space all their own, and separating them from their past and future relations with other forms of land use (such as maize or rice), fallow land emerges as the "empty" containers of low economic value and high carbon potential that make them ideal for a carbon offset (cf Bassett and Zuéli 2000).

I describe this process in order to draw attention to how this understanding of agricultural space is situated within a wider discursive formation in which the indigenous body, cacao, and indigenous agriculture more generally are spoken about in particular ways. Similar to the way in which cacao agroforestry is spoken about as a way of optimizing space, each form of land use is posited as a container of value, where the project offers a way to "fill" these spaces with potentially valuable carbon-sequestering biomass. For these calculations to have meaning, however, one must assume that the spaces are "managed" by the rational maximizer of neoclassical economics. This can be read as the application of universal economic theories and assumptions to a local context, a practice that has come under scrutiny of a number of critics of neoliberalism and neoclassical economics more generally (eg Peck 2004; Robertson 2006). I posit, however, that such calculations can only have meaning within a discursive formation where the discursive "work" that makes such a calculation understandable, even possible, in this context has been ongoing for quite some time. In this case, the previous emergence of the "indigenous land manager" as a rational economic agent—the one who "maximizes economic efficiency" of Proyecto NAMASÖL—was a necessary precondition for such calculations to be taken seriously within a context of agricultural development in Talamanca.

In other words, the ability of the carbon calculations to emerge as discursive statements that can be *evaluated as* true or false comes from their relation to a historically embedded set of discursive rules—rules that allow for these calculations to discursively relate indigenous bodies to agricultural space in these particular ways. As Foucault argues, no statement can exist in isolation, but is always understood in relation to a field of similar statements (see Foucault 1972:99–100). Understood this way, these calculations are more than a neutral evaluative tool that was diffused from the regulatory structure of the Kyoto Protocol or the discipline of neoclassical economics. These calculations are also embedded within, and transformative of, past conceptions of the indigenous body, and its relation to agricultural space. Past discursive statements posited the indigenous body as an economically rational manager of discrete agricultural spaces, ultimately allowing for the intelligibility of a cost–benefit calculation where *rastrojos* "open up" as discrete spaces of potential carbon value. The end result was a project that had to re-orient its original goal. Instead of promoting an expansion of cacao agroforestry, these cost–benefit calculations forced the project to look toward sites without an apparent economic value, with the rastrojos as a leading contender for receiving carbon. The result is a final project plan that calls for 30% of the project's carbon sequestration to come from eliminating *rastrojos*, 29% to come from adding additional trees to current banana plots, 26% from additional trees to cacao plots, and 15% to come from reforesting riverbanks (Segura 2005:80).[5]

The Value of Additionality and the Discourse of Value

While the calculations can be seen as part of a discursive formation that helped give rise to a development problematic, I contend that an engagement with value theory is needed to understand why such calculations are needed at all. Using a Marxian approach toward value, I argue that the practices of calculation and quantification at the point of a carbon offset's production are necessary discursive statements for the carbon forestry offset to acquire a use value. I make this argument by drawing on the importance of a commodity's use value at the point of what Marx refers to as the *salto mortale* of value (Marx 1976 [1867]:201)—the moment after a commodity is produced, but has not yet been sold, and must make the fatal leap from production to consumption for value to be realized.

In order to show how practices of calculation and measurement are sources of a carbon offset's use value, I draw on the insights of Kojin Karatani (2003), and his argument that Marx showed that it is only *after* the commodity is sold that the value created in the production process is *realized* [*verwirklicht*], and it is only from this *ex post facto* perspective that one can see the commodity's form as a synthesis of both use and exchange value:

> A certain thing—no matter how much labor time is required to make it—has no value if not sold...Classical economists believe that a commodity is a synthesis between use value and exchange value. But this is only an ex post facto recognition. Lurking behind this synthesis as event is a "fatal leap (salto mortale)" (Karatani 2003:8).

The *salto mortale* that Marx describes in *Capital*, Vol I (Marx 1976 [1867]:201) is the moment when the capitalist puts a commodity into the exchange relation—when the commodity object enters into an equivalence relation with money. It is in this moment of the *salto mortale* that the use value of a commodity takes on a special importance. This is because the production of value at the *point of production* is merely the production of *potential* values, and the critical synthesis between use value and value that embodies the commodity form does not emerge until the commodity crosses the threshold from production to exchange, and the commodity is purchased and the use value of this object is realized. Marx writes:

> All commodities are non-use-values for their owners, and use-values for their non-owners. Consequently, they must all change hands. But changing of hands constitutes their exchange, and their exchange puts them in relation with each other as values and realizes them as values. Hence commodities must be realized as values *before* they can be realized as use-values (Marx 1976 [1867]:179, italics mine).

And yet paradoxically, this exchange (C-M′), where money is advanced and a commodity becomes a use value (and then also results in value for the producer) cannot happen unless the commodity *already has* a use value before the exchange occurs. Marx continues:

> On the other hand, they [commodities] must stand the test as use values before they can be realized as values. For the labor expended on them only counts in so far as it is expended in a form which is useful for others. However, *only the act of exchange* can prove whether that labor is useful for others, and its product consequently capable of satisfying the needs of others (Marx 1976 [1867]:179–180, italics mine).

In other words, the producer of a commodity must produce use values for others for their commodities to have value; however, a commodity only *becomes* a use value after it is sold and is useful for the consumer of a commodity. The producer must accept a leap of faith that the commodity will have a use value for someone and the value congealed in the commodity (measured in abstract socially necessary labor time) may be realized.

Though abstract, Marx's analysis of the value form shows how additionality calculations enable a carbon offset to have a use value. In theory, the usefulness of a carbon offset is to allow a person or industry to emit carbon dioxide (and other greenhouse gases) in a way that does not adversely impact the climate. Additionality calculations are needed to ensure that an offset consumer knows that the project they are financing results in additional carbon in the ground, rendering his equivalent emissions as "climate neutral". Under the framework of the CDM, a forestry offset is dependent upon its demonstration that it "really is" contributing to this worldwide mitigation of carbon. If money advanced for a carbon offset were subsidizing already-existing trees, an offset would fail to neutralize someone's greenhouse gas emissions, and would not be useful in this sense. To avoid this, a forestry offset must demonstrate that the space of an offset project will one day be occupied by a complex system of trees, soils, hydrology, and so forth, and that this assemblage of carbon-sequestering biomass will be there *because of* the sale of a carbon offset credit. In other words, an offset must be put in an exchange relation before it can become useful for someone; however, this offset must be *shown to be useful* before it can be exchanged. Here, additionality calculations fill this role. Through these calculations, the commodity's use value is demonstrated, the *salto mortale* of exchange can be completed, and value can be realized.

I contend that an understanding of use value from the perspective of the salto mortale can explain why the carbon forestry offset assumed the form that it did. Seen from the perspective of consumption, a carbon offset requires that a consumer calculate its carbon relationship with the world, where one determines a quantified level of carbon dioxide that needs to be "offset". By quantifying one's climatic impact, a specific carbon-emitting action in one location can now be made commensurate with a level of carbon stored in the ground somewhere else. In this way, the quantified relation between a potential offset consumer and her carbon dioxide emissions allows for the carbon storing capacities of a forestry offset to become useful. For example, a potential offset consumer in Belgium may have a vague idea that her factory's emissions are contributing to climate change, however, the carbon sequestering properties of a *rastrojo* field in Costa Rica are of no use to that owner until her factory is put into a relation with a global regime of climate regulation (ie the Kyoto Protocol). Thanks to emissions cuts that are mandated by Kyoto, the carbon dioxide externalities of this owner's factory have now been quantified and found to be above the regulatory limit. Now the owner has costly emissions reductions she must meet. By quantifying her factory's emissions, and by putting these emissions in relation to a global management of the atmosphere, the carbon that is sequestered in an abandoned *rastrojo* field has now become useful for the factory owner. In other words, it is through a consumer's quantified relation to a global climate management regime that a particular ordering of carbon is needed, and it is through this coordinated ordering of carbon that a carbon offset project in Costa Rica becomes useful to someone on the other side of the planet.

Such usefulness, however, cannot help overcome the *salto mortale* of value until similar calculations are performed on the production end, and offsets are demonstrated to be useful. Simply put, a CDM offset's usefulness within the Kyoto Protocol is centered on its contribution to a worldwide coordination of the global carbon cycle, where carbon dioxide emissions in one place can have, in theory, a neutral climatic impact due to an equivalent level of carbon sequestration that occurs somewhere else. For such a geographically dispersed management of the climate to occur, however, various additionality calculations are necessary so that offset consumers can be assured that money advanced from a credit sale will materialize in new forms of carbon storage. Such calculations provide assurances that forestry offsets will help, for example, a corporation or nation meet the requirements of the Kyoto Protocol. And for the factory owner in Belgium, offsets are only useful to her to the extent that they help her meet her emissions requirements. Without calculations that demonstrate an offset's additionality, it is not useful in either sense.

In other words, under the precarious choreography of the global carbon cycle that has come to define the Kyoto Protocol, it is not the carbon-in-the-ground that gives an offset its use value. Instead, it is the relational ordering between the spaces of carbon storage, the carbon dioxide emitter, and the atmosphere itself that ultimately makes a forestry offset useful. Furthermore, I contend that in the context of this management of carbon, these practices of calculation do not merely *represent* the ordering of carbon that occurs within this framework, but rather, the calculations themselves and the relational ordering of carbon become effectively inseparable. This inseparability can be seen through the function of the concept of additionality. Under the Kyoto Protocol, it is only the carbon forestry offset's *demonstrated* additionality that counts towards that project's value (Chomitz 2000). A particular project may be *de facto* additional, but if it is not demonstrated to be such, it cannot be useful to the factory owner trying to meet particular emissions reduction standards. Thus, additionality calculations are more than abstractions that *represent* an offset's use value, but rather, the calculations are needed to demonstrate a project's usefulness, and allow for the *salto mortale* of value to be completed. Thus, under the overdetermined framework of carbon offset trading, an offset's materialization becomes inseparable from its representations, and the calculations themselves become the useful thing.

This folding of quantification into use value results in particular discursive transformations of space. To conduct a cost–benefit calculation that would show additionality, each space was treated as a discrete space of carbon-value potential. To do so, the indigenous body's relation to space becomes discursively transformed. No longer is he managing a complex, interrelated series of agricultural spaces, but instead, atomized containers of carbon value potential. Here, the spaces of the *rastrojo* have emerged not as a hinge that links other agricultural spaces (as with the Proyecto NAMASÖL report), but instead, as a valueless space of carbon sequestration potential. While this was a process that was conditioned by previous discursive formations of the indigenous body and space, it also transformed these objects in new ways through framings of space that are necessitated by the calculatory demands of creating value in a carbon offset.

Conclusion

In this paper, I have analyzed how specific spaces become opened up as sites of commodified carbon storage by treating the cost–benefit calculations of a carbon forestry offset as a discursive statement that is needed for this commodity to have value. Doing so has led me to argue that the intelligibility and significance of these calculations derive, in part, from their connections with other statements within a wider discursive formation of agricultural development. In the process, I have demonstrated that their impact as statements derive from their embeddedness within a history of development interventions in this area, from which a sedimented discourse about the indigenous body and its relation to agricultural space has emerged. In other words, these calculations are able to occur, and be taken seriously, through the emergence of the indigenous body, and indigenous agriculture, as discursive objects that are spoken about in particular ways. Such a "local" discursive formation helps to explain how carbon offsets were able to emerge as the solution to the problem of indigenous agriculture.

Understanding this context, however, does not explain why such calculations need to occur. Thus, I have also analyzed these calculations in their role as facilitating the production of value. I have argued that the calculations themselves are the offset's use value. This is a condition that derives from the paradoxical and uncertain position a commodity is in before it is sold, and value is realized, where the commodity is not realized as a use value until this moment, but yet must still be a use value before the *salto mortale* of value occurs. I have demonstrated that, when confronted with the exigencies of producing a commodity with use value in the context of the Kyoto Protocol, the indigenous body, and its relation to agricultural space, becomes discursively transformed in significant ways, allowing for particular agricultural spaces to be seen as "ready" for the production of carbon value, and foreclosing on other possibilities.

A number of writers have shown the difficulties of commodifying nature, difficulties that emerge when the accumulation and circulation demands of capital are confronted with the materiality of the non-human's physical properties. This uneasy marriage can mean that some natures are "uncooperative" (Bakker 2003) and extremely difficult to fully commodify, or that other natures need to be understood in radically simplified ways so they can be "read" by capital (Robertson 2006). With this case, I extend these insights by showing how the properties of particular natures come to be desirable as commodities in the first place, and how the exigencies of value condition how some natures and spaces can become commodities. As Castree (2008a) suggests, such a case can be seen as a type of environmental fix, one that is not immune to the problems that other writers on nature's commodification have explicated (eg Bakker 2003; Boyd 2009; Robertson 2006). I show here, however, that the question of why some spaces become commodified cannot be reduced to the global process of capital's contradictions alone, even if the demands of realizing value ultimately condition how a project may unfold. Instead, my approach seeks to problematize the relation between the global and the local in the production of a commodity. In this case, the project is simultaneously situated within a "local" development context—a position by which the project has become desirable—and a global project of regulating a worldwide carbon cycle. This is a position that, when combined with the requirements of value, requires specific quantified representations of the spaces of an offset, and ultimately conditions the kinds of spaces that are available for carbon commodification. In this case, *rastrojos* became a site of commodification through a process by which this space came to be understood as a "valueless" space of carbon potential.

Thus, my aim in this paper has not been to show how globally emergent "neoliberal" processes such as commodification become applied to a local context. Instead, my goal has been to show how a particular discursive formation allows for specific neoliberal interventions to arise (cf Boyd 2009). By evaluating the cost–benefit calculations as discursive statements that enable the production of value, however, I have shown that the properties of the commodity form require an opening-up of different spaces entirely. In this case, local complexity did not alter the implementation of abstract neoliberal ideas, but rather the process of value conditioned the manner in which this specific project was able to unfold. This is particularly significant because this was not a case where calculations were done to provide a commensurability needed for exchange, but rather, a quantified commensurability between producers, consumers, and the global climate was required for the offset commodity to become useful, and for value to be realized.

Acknowledgements

With the usual disclaimers, I thank Kendra McSweeney, Joel Wainwright, Becky Mansfield, Kevin Cox, and three anonymous referees for their comments on previous drafts of this paper. In addition, I am grateful to the project staff at CATIE for their openness, goodwill, and assistance. Fieldwork for this paper was funded with a dissertation fellowship from the Social Science Research Council and American Council of Learned Societies.

Endnotes

[1] The agricultural extension component reached approximately 64 different households (Bodegom, Sanders and Brenes Castillo 2000). This is smaller than agroforestry projects undertaken by CATIE (described later), which usually involve anywhere from 100 to 500 farmers. To my knowledge, no systematic study of the long-term success of these projects has been undertaken. However, field observations, anecdotal evidence, and the fact that very similar agroforestry improvement projects have been

ongoing since the early 1980s suggest that the long-term efficacy of these projects is questionable.

[2] While the report that informs the NAMASÖL project is remarkably consistent with other development projects, I note here that the goals of the NAMASÖL project differed from the projects carried out by CATIE (described later) in important ways. The external practitioners were different even if local indigenous liaisons and participants were the essentially the same. In addition, agriculture was only one component of a broader capacity-building focus of NAMASÖL while CATIE's projects were more centrally focused on promoting sustainable agriculture.

[3] One of the authors of this report, Carlos Borge, is frequently employed to write consultancy documents for other development and state agencies working in this region, and has contributed to consultancy reports for a number of CATIE's projects as well.

[4] Cacao cultivation still exists in Talamanca, albeit at a drastically reduced scale. One of primary challenges for households to switch from plantain to cacao is the four to five year lag time it takes for newly planted cacao trees to begin bearing fruit.

[5] Currently, the project is stalled in the development stage because of lack of further World Bank funding for project implementation. Project managers were currently looking to turn this project into a voluntary offset, but at the time of publication, have yet to do so successfully (anonymous interview 2008).

References

Agrawal A (2005) *Environmentality: Technologies of Government and the Making of Subjects*. Durham, NC: Duke University Press.

Althusser L (1979 [1965]) *For Marx*. London: Verso

Bakker K (2003) *An Uncooperative Commodity: Privatizing water in England and Wales* Oxford: Oxford University Press

Bassett T and Zéuli KB (2000) Environmental discourses and the Ivorian Savanna. *Annals of the Association of American Geographers* 91(1):67–95

Beer J (1991) Implementing on-farm agroforestry research: Lessons learned in Talamanca, Costa Rica. *Agroforestry Systems* 15(2–3):229–243

Birkenholtz T (2009) Groundwater governmentality: Hegemony and technologies of resistance in Rajasthan's (India) groundwater governance. *Geographical Journal* 175(3):208–220

Bodegom A, Sanders A and Brenes Castillo C (2000) *Informe de Evaluación de Proyecto Namasöl* [Project Namasöl evaluation report]. San José, Costa Rica: Embajada de los Países Bajos

Borge C and Castillo R (1997) *Cultura y Conservacion en la Talamanca Indigena* [Culture and conservation in indigenous Talamanca]. San José, Costa Rica: Universidad Estatal a Distancia

Borge C and Laforge M (1996) *Estrategia de Transferencia de Tecnologia en Talamanca* [Strategy for technology transfer in Talamanca]. Consultancy report for Proyecto NAMASÖL, San José, Costa Rica: Museo Nacional

Borge C and Villalobos V (1995) *Talamanca en la Encrucijada* [Talamanca at the Crossroads]. San José, Costa Rica: Universidad Estatal a Distancia

Boyd E (2009) Governing the Clean Development Mechanism: Global rhetoric versus local realities in carbon sequestration projects. *Environment and Planning A* 41(10):2380–2395

Brown K and Corbera E (2003) Exploring equity and sustainable development in the new carbon economy. *Climate Policy* 3S1:S41–S56

Bumpus A G and Liverman D M (2008) Accumulation by decarbonization and the governance of carbon offsets. *Economic Geography* 84(2):127–155

Castillo R (1999) The expansion of plantain monoculture in the Talamanca Indian Reserve, Costa Rica. *Mesoamerica* 4(2):69–74

Castree N (2003) Commodifying what nature? *Progress in Human Geography* 27(3):273–297

Castree N (2008a) Neoliberalising nature: The logics of deregulation and reregulation. *Environment and Planning A* 40(1):131–152

Castree N (2008b) Neoliberalising nature: Processes, effects, and evaluations. *Environment and Planning A* 40(1):153–173

CATIE (1995) Institutional development plan. Unpublished document. Turrialba, Costa Rica: CATIE

CATIE (2006) "Proyecto Fijacion de carbono en sistemas agroforestales con cacao de fincas indigenas entalamanca, Limon, Costa Rica." [Carbon capture project in agroforestry systems with cacao in indigenous farms in Talamanca, Limon, Costa Rica] Second phase project proposal for *Proyecto Captura de Carbono y Desarrollo de Mercados Economicos en Sistemas Agroforestales Indigenas con Cacao en Costa Rica, TF-052118* [Carbon capture project and development of economic markets for indigenous agroforestry systems with cacao in Costa Rica]. Turrialba, Costa Rica: CATIE

Chomitz K M (2000) *Evaluating Carbon Offsets from Forestry and Energy Projects. How Do They Compare?* Policy Research Working Paper # 2357. The World Bank, Development Research Group, Infrastructure and Environment

Dahlquist R M, Whelan M P, Winowiecki L, Polidoro B, Candela S, Harvey C A, Wulfhorst J D, McDaniel P A and Bosque-Perez N A (2007) Incorporating livelihoods in biodiversity conservation: A case study of cacao agroforestry systems in Talamanca, Costa Rica. *Biodiversity and Conservation* 16(8):2311–2333

Ferguson J (1990) *The Anti-Politics Machine.* Cambridge: Cambridge University Press

Foucault M (1972) *The Archaeology of Knowledge and the Discourse on Language.* New York: Pantheon Books

Gidwani V (2008) *Capital Interrupted.* Minneapolis: University of Minnesota Press.

Guzmán J (2006) Evaluación de Resultados e Impactos Económicos [Evaluation of results and economic impacts]. Final report for Carbon Capture and Development of Environmental Service Markets Project TF-052119. Turrialba, Costa Rica: CATIE

Harvey C A, Gonzalez J and Somarriba E (2006) Dung beetle and terrestrial mammal diversity in forests, indigenous agroforestry systems and plantain monocultures in Talamanca, Costa Rica. *Biodiversity and Conservation* 15(2):555–585

Hinojosa Sardan V R (2002) "Comercialización y certificación de cacao (Theobroma cacao Linn.) y banano (Musa AAA) orgánico de las comunidades indígenas de Talamanca, Costa Rica" [Commercialization and certification of organic cacao and banana from indigenous communities in Talamanca, Costa Rica] Unpublished Master's thesis, CATIE, Turrialba, Costa Rica

Karatani K (2003) *Transcritique:On Kant and Marx.* Cambridge, MA: MIT Press

Marx K (1976 [1867]) *Capital: A Critique of Political Economy*, Vol I. New York: Penguin

MacKenzie D (2009) Making things the same: Gases, emission rights and the politics of carbon markets. *Accounting, Organizations and Society* 34(3–4):440–455

Michaelowa A (2005) Determination of baselines and additionality for the CDM: A crucial element of credibility of the climate regime. In F Yamin (ed) *Climate Change and Carbon Markets: A Handbook for Emissions Reduction Mechanisms* (pp 305–320). London: Earthscan

Nair P K R (1993) *An Introduction to Agroforestry.* Dordrecht, The Netherlands: Kluwer

Peck J (2004) Geography and public policy: Constructions of neoliberalism. *Progress in Human Geography* 28:392–405

Polidoro B A, Dahlquist R M, Castillo L E, Morra M J, Somarriba E and Bosque-Perez N A (2008) Pesticide application practices, pest knowledge, and cost–benefits of plantain production in the Bribri-Cabécar Indigenous Territories, Costa Rica. *Environmental Research* 108(3):98–106

Robertson M (2006) The nature that capital can see: Science, state, and market in the commodification of ecosystem services. *Environment and Planning D: Society and Space* 24(3):367–387

Schroeder R A (1999) *Shady Practices: Agroforestry and Gender Politics in The Gambia.* Berkeley, CA: University of California Press

Segura M (2005) Estimación del carbono almacenado y fijado en sistemas agroforestales indígenas con cacao en la zona de Talamanca, Costa Rica [Estimation of stored fixed carbon in indigenous cacao agroforestry systems in the Talamanca zone, Costa Rica]. Unpublished technical document for Proyecto Captura de Carbono y Desarrollo de Mercados Ambientales en Sistemas Agroforestales Indígenas con Cacao en Costa Rica, TF-052118

Somarriba E (1981) "Sistema Taungya:Tecnología Apropiada de Repoblación Forestal (Revision de Literatura) [Taungya system: Appropriate technology for forest repopulation (review of the literature)]." Unpublished manuscript. Turrialba, Costa Rica: CATIE

Somarriba E (1993) "Cacao-plátano-madera: La diversificación agroforestal como herramienta para manejar variabilidad en precios de productos agricolas" [Cacao–plantain–wood: Agroforestry diversification as a tool for managing price fluctuations in agricultural products]. *CATIE, Turrialba. Programa de Agricultura Tropical Sostenible Semana Científica. Memorias.* Turrialba, Costa Rica: CATIE

Somarriba E (1997) *Cacao Bajo Sombra de Leguminosas en Talamanca, Costa Rica, Manejo, Fenología, Sombra y Producción de Cacao* Serie Técnica no 289 [Cacao under leguminous shade in Talamanca, Costa Rica, Management, phenology shade and cacao production]. Turrialba, Costa Rica: CATIE

Somarriba E and Beer J (1999) Sistemas agroforestales con cacao en Costa Rica y Panama [Agroforestry systems with cacao in Costa Rica and Panama]. *Agroforesteria en las Américas* 6:7–11

Somarriba E, Dominguez L and Lucas C (1994) *Generación y Transferencia de Tecnología. Proyecto Agroforestal CATIE/GTZ: No. 6. Serie Técnica no 233* [Technology generation and transfer. Agroforestry project CATIE/GTA: No 6. Technical series no. 233]. Turrialba, Costa Rica: CATIE

Somarriba E, Villalobos M, Gonzalez J and Harvey C A (2004) *El proyecto conservación de biodiversidad y producción sostenible en pequenas fincas indígenas de cacao orgánico en el corredor biológico Talamanca–Caribe, Costa Rica* [Project for biodiversity conservation and sustainable production in small, indigenous organic cacao farms in the biological corridor Talamanca-Caribe, Costa Rica]. Semana Científica 2004, Memoria. Turrialba, Costa Rica: CATIE

Vargas Carranza J (1985) "Dinámica de la Ocupación Territorial y formación especial del Grupo Bribri, Valle de Talamanca, Costa Rica [Dynamic of the territorial occupation and formation of the Bribri Group, Talamanca Valley, Costa Rica]." Unpublished Master's thesis, Geography Department, University of Costa Rica, San José

Wainwright J (2008) *Decolonizing Development: Colonial Power and the Maya.* Oxford: Blackwell

Chapter 8
Resisting and Reconciling Big Wind: Middle Landscape Politics in the New American West

Roopali Phadke

Introduction

On the edge of the Mojave Desert in the western USA, just southeast of Las Vegas' glittering casino marquees, a battle is brewing over the siting of Nevada's first utility-scale wind project. Duke Energy Corporation proposes to build a US$600 million energy facility consisting of 140 turbines to generate 370 MW of electricity.[1] The project is located near the community of Searchlight, Nevada. This historic mining town is surrounded by protected public lands teeming with threatened and endangered species, including desert tortoise, gila monsters and bighorn sheep. Searchlight is also the birthplace and home of the Majority Leader of the US Senate, Harry Reid, an ardent supporter of increasing America's renewable energy capacity.

The Searchlight project is an important test case for the development of wind energy in the American West. The federal Bureau of Land Management (BLM) is overseeing the permission for this project. The environmental review is at the very beginning stages with a record of decision anticipated in 2011. While local opposition to the Searchlight project is still crystallizing, over 100 people showed up to the BLM's first public "scoping" meetings. Local residents expressed broad concerns about the impacts of industrial, "big" wind development on sensitive species habitat, archaeological ruins, and scenic corridors (BLM 2009). While it was once rare that local citizens would organize opposition to

The New Carbon Economy, First Edition. Edited by Peter Newell, Max Boykoff and Emily Boyd.
© 2012 Roopali Phadke. Book compilation © 2012 Editorial Board of Antipode and Blackwell Publishing Ltd.

utility-scale wind projects this early in the permitting cycle, it is now increasingly the norm throughout the country.

America's western landscapes have long been associated with productive energy development. To the east of Las Vegas, Hoover dam's massive turbines have spun water into electricity since the 1930s. More recently, a national nuclear waste depository has been proposed at Yucca Mountain. Nevada's remoteness may suggest that it is an ideal place for new clean energy development, particularly wind and solar facilities. Yet, Americans also strongly associate Nevada with a protectionist ideal. Nature enthusiasts cherish the opportunity to recreate and commune in these wild landscapes.

As local opposition to wind energy continues to grow, this is an important time to examine how government officials, developers and citizens are balancing the protection and development of wild and rural landscapes in the context of a new global low carbon energy economy. This article examines the nascent US wind energy opposition movement as evidence of emerging social resistance to the re-sculpting of energy geographies. My characterization of this emerging opposition movement focuses on the shifting politics of the American West. I draw on cultural landscape and place theory to understand how oppositional campaigns, including the one in Searchlight, call for reconciling the "middle landscape" ideal with strong social and political commitments to a new wind energy economy. The goal of this paper is to challenge the presumption that such oppositional movements are simply examples of "not in my backyard" (NIMBY) politics. Instead, I argue that local opposition to wind development in the "New American West" is representative of broader shifts in the economic and aesthetic value of once historically "productive" rural landscapes. Accusations of NIMBYism tend to sideline important concerns about the pace, scale and potential rural economic benefits of wind energy development.

First coined by Leo Marx in his seminal 1964 book *The Machine in the Garden*, the "middle landscape" referred to a near impossible reconciliation of modern technology with an American pastoral ideal. Marx wrote that the "middle landscape" embodied a deep-seated American contradiction: an attempt to define a nation's purpose as both the "pursuit of rural happiness while devoting itself to productivity, wealth and power" (1964:226). As a conspicuous technology capable of both sustaining and thwarting the realization of an imagined rural ideal, utility-scale wind energy development across American landscapes calls forth a similar social reconciliation as was faced with the expansion of railroads, steamships and factories in early modern America. Contemporary struggles over the landscape impacts of rural "green" industrialization projects are complicated by rapid shifts in rural demographics, economics and cultural values.

The article is divided into four main parts, excluding this introduction. The first section integrates literatures from cultural landscape and place theory with the emerging "New West" scholarship that critically examines the shifting demographics, economics and environmental values taking hold in regions of in-migration in the American West. The second part describes public perceptions of wind energy and the characteristics of the nascent wind opposition movement. The third section develops a case study of the Searchlight project as a representative example of New West opposition. The fourth, and final, section argues for the need to chart a new course for wind development where local residents retain a degree of sovereignty over landscape values and imaginaries. Drawing on Leo Marx's scholarship, I suggest ways that new energy targets must be negotiated in the contexts of powerful social commitments to a "middle landscape" ideal.

The material presented in this article is the result of multiple qualitative research methods. These include semi-structured interviews with government officials, wind energy developers, movement leaders and local residents living near proposed wind projects. Two years of observational studies have been conducted at national and regional wind industry conferences. A range of case studies have also been researched and written about contentious projects across the nation.[2] My research on the Searchlight case has included a field visit and personal interviews with local activists and federal agency staff. Lastly, GIS mapping tools were used to plot and analyze the growth of this opposition movement.

Mobilizing Place Theory and New American West Literatures

Wind energy opposition politics are essentially battles over rural space; over who controls the productive and consumptive qualities of rural landscapes. North American geographic scholarship has been slow to respond to these emerging politics. I critically engage cultural geography and place theory to better understand the characteristics of the nascent wind energy opposition movement. Toward this goal, I weave together several strands of "place-based" scholarship. I draw from Leo Marx's insights in *The Machine in the Garden* as a way of connecting historical and modern concerns with technology and landscape. I am also interested in hybridizing two currents of landscape theory that are particularly germane to wind energy politics. On the one hand, cultural anthropologists and political ecologists celebrate place-based struggles as defensive challenges to globalization and the commodification of culture. On the other hand, cultural geographers have argued that such "aestheticized" movements can in fact represent dangerous and reactionary retreats into an unrepresentative, elite politics of the local. I engage the New West literature to situate place politics within a broader set of rural trends that demonstrate important shifts in meanings around the productive and consumptive ethics and economics of landscape aesthetics.

Cultural geographers and cultural anthropologists have been fascinated by the links between landscape identity and place attachment for the last four decades. In his 1974 book, Yi-Fu Tuan coined the term "topophilia" to represent the "affective bond between people and place or setting" (1974:4). Tuan's early work connected topophilia with Leo Marx's concept of the "middle landscape", which Tuan described as "the ideal middle world of man poised between the polarities of city and wilderness" (1974:109). In *The Machine in the Garden*, Leo Marx traced the importance of "pastoralism" as a distinct American social, literary and political form. He argued that the psychic root of all pastoralism was a yearning for a simpler existence "closer to nature" (1964:6). The "middle landscape" offered a place "secluded from the world—a peaceful, lovely, classless, bountiful pasture" (1964:116).

The "middle landscape" was not a static state but a dialectical image of society in constant need of redefinition. The early American rhetoric of progress exalted technology, embodied by the "simple, stark power" of steamships, millwheels and railroads, as an agent of social change that helped realize the abundance of the "middle landscape" ideal. The shear industrial transformation of the American landscape that resulted from the technological changes of the nineteenth century did little to displace this notion of pastoralism. Marx wrote that "the pastoral ideal remained of service long after the machine's appearance in the landscape" (1964:226). Both Tuan and Marx have had a direct influence on the more recent "sense of place" literature, which argues that enduring places result from strong, participatory community development processes.

The cultural anthropology and geography work on this subject has developed a neo-Marxist critique about the geographies of struggle, resistance and social control implicated in place making. In particular, the political ecology scholarship on place sees the "middle landscape" as a site of constant struggle and negotiation. Arturo Escobar has argued that the politics of place are often about the emergence of new identities within region-territories. Escobar's scholarship has examined the creative ways in which social movements produce a politics of "defensive localization" (2001). Chronicling the "subaltern strategies of localization" operationalized by communities, he argues that social movements employ concepts of biodiversity and sustainability as part of their cultural struggles for political autonomy and self-determination. These movements aim to counter state development policies that increase the pace of industrial extraction at the regional scale (Escobar 2001). While Escobar's examples are situated in the developing world, "First World" political ecologists extend these concerns to North America by claiming that culture and identity remain central to constructions of environmental politics (McCarthy 2002; Schroeder and Albert 2006).

In comparison to the above cited social movement scholarship that sees place politics as a destabalizing force on global capitalism, cultural geographers have also argued that celebrations of place can represent retreats into a form of localism that is more "widespread and insidious than is often acknowledged" (Duncan and Duncan 2004:26). In their 2004 volume, *Landscapes of Privilege*, Duncan and Duncan examine the social and political consequences that flow from various readings of the suburban landscape in Bedford, New York. They write that a seemingly "innocent appreciation of landscape and desire to protect local history and nature can act as a subtle but highly effective mechanism for exclusion and reaffirmation of class identity" (2004:4). Furthering this point, they cite David Harvey's claim that landscape debates often result in "the reactionary politics of an aestheticized spatiality" (2004:26).

Geographers focused on the "New American West" have also examined contestations over landscape ideals in the context of rural communities stuck at the crossroads of traditional resource-based production and new economies of landscape consumption. Peter Walker and Louise Fortmann's study of landscape politics in Nevada County, California stressed that new conflicts are emerging in places in the global North where "economic and cultural value is being placed not on individual natural resources but on aesthetic and environmental values (such as 'viewshed' or 'rural quality') that derive from a totality of many individual landholdings" (2003:471). They tell us that rural landscape politics revolve around the question of who "owns" the landscape and who decides how it "should" look.

Walker and Fortmann's work is part of a new dynamic inter-disciplinary articulation of the changing landscape politics of the American West. As Robbins et al argue in their review article, "New West research is predicated on well-established demographic and economic trends, with much emphasis on socioeconomic differences (and conflicts) between amenity-based communities and long-standing productivity-based ones" (2009:360). They suggest that new migrants, including the rise of those with "footloose" incomes and service sector jobs, bring new visions about the balancing of environmental priorities and economic development values. While residents were traditionally employed in extractive sectors, like logging, mining, ranching and farming, new migrants are drawn to the West's "quality-of-life" features. Robbins et al suggest the need for more research that examines how extractive development and amenity economies can be "smoothly" combined.

While the New West literature pays attention to how new migrants tilt rural priorities toward "green" thinking or "greentrificatrion", there are a spectrum of environmental values that get activated in the sociai construction of place. The 2010 Land Institute study on climate change planning in the American West suggested that local residents and planners do not name climate change as one of the most pressing planning challenges facing their communities. In comparison, concerns about wildlife protection and open space conservation rank high in the region. It is quite striking that the New West literature has not paid significant attention to how "industrial" scale wind energy development is creating even further tensions between extractive/ productive development and consumption based amenity economies. I raise this not as a critique of the scholarship, but rather to point to how the rapid pace and scale of wind development in the West has essentially caught both scholars and planners off guard. While wind developers and federal agencies have proposed thousands of new megawatts of installed projects, local government officials and residents are still coming to terms with the regional transformations these projects will engender.

Drawing on the above cited literatures, this article examines how the local wind energy opposition movements are resisting a national charge to re-sculpt America's rural energy geographies in the interest of climate change mitigation. While my analysis accepts Duncan and Duncan's claims that landscape preservation interests can exhibit "subtle naturalizing and aestheticizing attitudes", I resist reading landscape protection as the work of only privileged groups interested in building exclusionary communities (Duncan and Duncan 2004:30). Rather, my analysis examines American wind opposition movements as performing acts of "defensive localization" that challenge a technological path of progress steeped in utilitarian values. Drawing on Leo Marx's theorizing of the "middle landscape", I argue that this social movement represents a dialectical struggle to maintain a rural landscape identity that reconciles a naturalized pastoral ideal with the sudden appearance of new "machines in the garden". There is a danger that NIMBY- based analyses of wind opposition will fail to seriously acknowledge the major landscape transformations under way with new energy development. The 400 ft tall turbines, and the transmission lines that will carry those electrons to population centers, will crisscross western horizons in ways that will transform the cultural and physical landscapes of "Big Sky Country".

Public Perceptions and Opposition to Wind Energy

While there have been studies that examine the emergence of wind oppositional movements in European and Australasian contexts, we know far less about the attributes of this nascent social movement in the USA and the impact it is having on "green" energy goals and rural politics (Barry, Ellis and Robinson 2008; Hindmarsh and Matthews 2008). The American wind opposition movement has grown right alongside the development of the industry. In 1980, the world's first wind farm, consisting of 20 turbines, was built on Crotched Mountain in southern New Hampshire. The Crotched project was a failure because of an overestimation of wind resources and turbines that frequently broke (Asmus 2000:61). The industry then single-mindedly moved west to California in the post-OPEC years as a result of favorable state policies. By 1995, California produced 30% of the entire world's wind electricity, through mammoth projects like Altamount Pass. The industry still carries a legacy of social fallout from the early California projects due to high bird mortalities and poorly designed aesthetic features.

From these initial steps, wind development is now accelerating in many regions of America. There are currently 35,000 MW of installed wind energy capacity in the USA. The American wind industry added an additional 10,000 MW of new generating capacity with Recovery Act stimulus incentives (American Wind Energy Association 2010). The Department of Energy's (DoE) ground breaking report "20% Wind Energy by 2030" argues that wind power capacity will need to reach over 300,000 MW over the next two decades to meet targets (2008:7). The DoE's plan relies on aggressive new development, particularly in the Intermountain West and Central Plains regions. Approximately 7000 out of the total 10,000 MW of new installed capacity added in 2009 occurred west of the Mississippi River. To date, most political efforts have focused on innovation pipelines and economic incentives. Far less attention has been paid to how local residents perceive wind energy's landscape impacts.

Having developed the most remote sites first, the next wind energy frontier will bring turbines in much closer proximity to the places where people live, work and recreate. As a result, oppositional groups have formed in most areas of active wind development, and areas where wind development is incipient, like Nevada. The opposition to the Searchlight project, which will be described in the following section, is evidence of what many wind policy scholars refer to as the "social gap" that exists between broad national public support for renewable energy development and vocal organized resistance to the development of concrete projects located near or adjacent to rural communities (Bell, Gray and Haggett 2005; Szarka 2004; Wolsink 2007). Table 1 lists a set of general factors that impact public perceptions of wind energy impacts.

The wind oppositional movement has grown dramatically across the USA over the last 5 years. Figure 1 depicts where community level

Table 1: Factors affecting public perceptions[8]

Physical	Turbine color, size, acoustics; farm size and shape; cumulative effects of neighboring projects; proximity to turbines and important landscape features
Political	Institutional capacity; public participation and consultation; perception of developer
Socioeconomic	Shareholding models; effect on property values; impacts to hosting communities
Symbolic	Representations of wind turbines
Local	Place and identity processes; local or community benefit and control
Personal	Previous experience and knowledge

groups are challenging wind energy projects in the USA. There are over 200 unique organizations plotted on this map. This map helps illustrate a few trends. First, most oppositional groups are located in the northeast where population densities and land values are among the highest in the nation. This region also has little installed capacity, in part due to the strength of the opposition. This mirrors media portrayals of wind opposition in the northeast, particularly the Cape Wind offshore project in Massachusetts, as driven by elite interests focused on protecting privileged "views" (Phadke 2010).

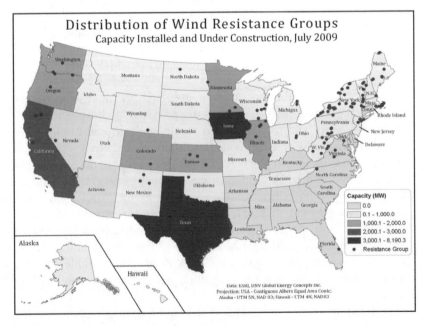

Figure 1: Community resistance to wind energy

Heightened controversies about wind siting in the northeast have helped drive development westward. While the Intermountain West has some of the best wind resources in the country, the westward movement of the industry has also been fueled by a presumption that wind energy will be more publicly acceptable in the Plains and Western states where there has historically been more industrial agriculture and energy intensive development (Sower 2006). Yet, Figure 1 also documents that there are wind opposition groups in these regions as well. As the Searchlight case will show, wind opposition is also quickly growing in regions where there is no installed wind energy capacity to date. This is particularly significant because the DoE estimates that the most rapid development over the next two decades will occur in states like Nevada, Montana and Wyoming, where residents have the least experience with wind energy on the landscape (2008).

Simple dots on a map tell us little about the characteristics of these "opposition groups". There is considerable range in size, budget and professionalism. Some organizations represent longstanding historic preservation and conservation interests with an established membership base. Yet, the majority of these organizations were created in response to a single project. Looking across these oppositional groups, it is possible to see the presence of a sense of shared rural identity in the face of wind development. The words "protect" or "preserve" appear in many organizational titles, such as Protect Pendleton, Protect the Flint Hills and the Ridgeprotectors of Vermont. As indicated in Table 2, wind opposition groups also frame their efforts in preservationist terms by using the word "save" in their organizational titles. Group websites highlight concerns that "industrial", "obscene", and "irresponsible" wind development is destroying rural landscapes and livelihoods. For

Table 2: Preservationist organizations

Save Burney's Skyline
Save Coteau Prairie Landscape
Save Crystal Lake
Save Our Allegheny Ridges (SOAR)
Save Our Scenic Area (SOSA)
Save Our Common Mountain Environment
Save Our Scenic Hill Country Environment
Save Our Sound
Save Our Unspoilt Landscape (SOUL)
Save Our Valley
Save St. Lucie Alliance
Save Up-State New York
Save Vermont Ridgelines
Save Western Ohio

example, the website of the North Texas Wind Resistance Alliance describes their region as a place of "quiet, rural character" being invaded by a secretive and powerful industry.

In the language of social movement scholars, these localized rural struggles often operate "rhizomically". Michael Woods' analyses of rural movements suggest that groups often share a common vocabulary and shared repertoire of tactics, but essentially exist in isolation from each other (Woods 2003:313). The wind opposition movement bears these characteristics. At the founding meeting of the American national wind opposition movement in May 2005, 40 people came together in Massachusetts from states as far away as Colorado and Kansas to talk about their concerns. Conference participants felt that there was a need for a network to coordinate US-based efforts and provide a library of resources. The group National Wind Watch was born out of this meeting and continues to serve as a loosely organized networking and information sharing hub for the movement.[3] Wind Watch website visitors and email listserv members receive daily news e-blasts and have access to an extensive online archive of images, videos, and documents. Yet, National Wind Watch has not organized into chapters that share a unifying structure.

The rapid growth of local opposition groups, and the emergence of a new federal political will to fast track wind development, has created a need to reorganize and recalibrate the goals of the movement. To date, wind opposition has been a reactionary movement undirected by any one single philosophy. The political affiliations and energy policy goals of these groups are exceptionally diverse. Some groups, such as Industrial Wind Action group, advocate for the development of wind power in some places, such as the Plains states, but not in the more densely populated northeast (DePillis 2007). Other groups, such as New Mexico Conservation and Renewable Energy Society and the Minnesota group, Clay County Citizens for Sustainable Energy, endorse sustainable energy goals but criticize a planning approach that has left local communities in the dark about project development.

The American wind opposition movement is moving through what social movement scholars call "cycles of protest" (della Porta and Diani 1999). Until recently response was reactively focused at the project site. Wind opposition groups mobilized local members to defeat or redesign a particular project in their community. However, more recently, political mobilization has gone beyond project protest toward broad code making practices that aim to shape energy policy in a moment of rapid regulatory change. For example, the Wisconsin group "Better Plan" was one of the first organizations to advocate wind ordinances as zoning codes that limit the scale of projects and their proximity to people. There are signs that the movement is gaining a more consolidated platform. National Wind

Watch has recently helped to organize the North American Platform Against Windpower (NAPAW) to connect groups in the USA, Canada, Mexico, the Caribbean and Central America (http://www.na-paw.org). This effort works in concert with the EPAW (European Platform Against Windpower), which has 350 signatory organizations from 19 countries.

The westward move of the American wind energy industry was in part catalyzed by a federal process to permit development on public lands. Since 2005, the Bureau of Land Management (BLM) has been the lead federal agency responsible for wind energy permission. Founded in 1946 to manage 264 million acres of public lands, the agency's mission has been oriented toward enabling the homesteading and settlement of the American West by managing public rangelands and providing leases, exploration, and production rights for commodities, such as coal, oil, gas, and sodium. The agency has the authority to process and grant right-of-way applications and authorizes wind energy site testing and development. Given this history, the BLM approaches wind energy development not as a new resource that requires new protocols for public engagement, but rather based on a 60-year old model of extractive energy development. The BLM public engagement protocols are based on the environmental impact review processes directed by the 1970 National Environmental Policy Act.

The federal process for siting and permitting wind energy projects influences the ways that the industry legitimates and accommodates public concerns. The environmental impact statement (EIS) process requires that developers assess wildlife, visual, noise and cultural impacts. Over the last 2 years, the wind energy industry has focused its efforts around two critical public concerns: wildlife and noise. The American Wind Energy Association, a national trade association representing wind power's associated industries, recently helped create the Wind Wildlife Institute to research, map and mitigate wildlife impacts. Likewise, more recent community opposition to noise impacts, particularly the effect known as "wind turbine syndrome", a contested illness brought on by exposure to low-frequency vibration, led the industry toward commissioning a third-party medical study on this topic (American Wind Energy Association and Canadian Wind Energy Association 2009). Yet, on the viewshed issue, which researchers have asserted is a prime concern motivating opposition, the industry has chosen to consistently argue that perceptions of "visual pollution" amount to intractable cultural differences that are impossible to objectively quantify. While developers at times accommodate public concerns by adjusting project configurations and colorizing turbines to gain public acceptance, this is usually the result of contentious environmental reviews and not part of a participatory technology strategy that incorporates public values at an early stage of project planning.

The general dismissal of viewshed concerns also demonstrates that the public's negative views of wind energy development are rarely analyzed as part of broader demographic and cultural shifts in rural economies. Rather, members of the wind industry, government agencies and the media often attribute community concerns to classic NIMBY self-interest driven parochialism (Kahn 2000). The most recent edition of the popular book *Wind Energy Basics*, long-time wind energy advocate and scholar, Paul Gipe, attributes NIMBY reactions to conventional class politics. He writes that NIMBY "is just another manifestation of trying to pass the social costs of energy choices—as in the Lincoln Navigator or the McMansion—on to other, and often less politically powerful, groups" (2009:156). In a March 2009 interview with the Associated Press, the Secretary of the Department of Interior, Ken Salazar, similarly stated that it was imperative that "we get this thing done and not get stuck in a not-in-my-backyard syndrome". He further argued that it was a "false choice" to pit aggressive development of renewable energy against the protection of the country's wildlife and treasured landscapes (Cappiello and Herbert 2009). These statements caricature wind energy opponents as selfish individuals who expect to have it "both ways": maintain a purified nature while still benefitting from the fruits of technological progress. While NIMBYism may be at play in certain cases and in individual reactions, discounting viewshed concerns as self-serving draws attention away from the large-scale transformations that wind energy brings to landscapes.

Industry leaders often claim that wind supporters are the "silent majority" in the communities where they work and that a few vocal opponents can overwhelm planning officials. In *Wind Power in View*, the authors assert "small numbers of dedicated opponents can and will attack projects, crushing developers with their passion" (Pasqualetti, Gipe and Righter 2002:37). Those residents who support wind power often see it as compatible with current agricultural land use patterns—hence the term "wind farm". Supporters include landowners who have enacted leases with wind companies so that they earn an annual return for hosting turbines. In addition to these landowners, wind power companies gain favor with residents by supporting local services, such as funding schools through "payments in lieu of taxes" agreements. The wind industry also claims that its investment in local communities helps produce "green jobs". The DoE's National Renewable Energy Lab estimates that 6–10 permanent new operations and maintenance jobs, and 100–200 seasonal construction jobs, are created for every 100 MW of wind energy installed (Smith 2007).[4] The bulk of new jobs generated by the wind industry will be in the manufacturing sector and not at local project sites.

Many wind policy researchers have aimed to complexify the assumption that community opposition can simply be equated with NIMBYism (Cass and Walker 2009; Devine-Wright 2005; Kempton et al 2005; van der Horst 2007). These authors contend that local opposition is often symptomatic of a "democratic deficit" in wind policy and project planning where local residents have little role in designing projects so that they are compatible with local landscape values (Hindmarsh and Matthews 2008). These studies claim that members of wind opposition movements are often not arguing in their individual self-interest, but rather evoking community concerns over landscape, memory and identity in their protests against wind projects.

Geographer Michael Woods has characterized emerging rural movements, like the wind opposition movement, as essentially struggles in the defense of a rural identity. Woods writes "In the new rural economy the commodification of rural landscape, culture and lifestyle is more important than the physical exploitation of rural land" (2003:312). Echoing this claim, opposition to wind development is often strongest in areas of exurban migration. In-migrants act to protect their financial and emotional investment by opposing developments that threaten the perceived rusticity of their new homes. Woods labels this form of politics as "aspirational ruralism", where residents resist developments that threaten an imagined pastoral ideal.

In areas of significant in-migration, wind developers are weary of what they refer to as the "Starbucks effect".[5] Reflected in Wood's "aspirational ruralism", this phenomenon refers to the presence of a "Starbucks" demographic of educated, affluent and political aware consumers who will likely block development they perceive threatens their imagined community. The wind policy scholarship provides an analytical lens into the Starbucks effect. van der Horst has written that "People who have moved to the countryside as a lifestyle choice and are less dependent on the traditional rural economy" are more likely to oppose wind projects (2007:2709). In the following section, I develop a case study of the Searchlight Wind project to illustrate these trends.

Searchlight Wind

Local opposition to the Searchlight Wind project in Nevada is representative of growing concerns about wind energy in the American West. Located 70 miles southwest of Las Vegas in Clark County, the small community of Searchlight has a population of just about 800 people (Census 2007). The town center consists of little more than a community center, two small casinos, a cemetery and school. While half of the population is retired, the other half is an eclectic mix of miners, ranchers, small business owners and artists. Searchlight's population grew by 39% from 2000 to 2010, largely due to an influx of new retirees. Yet, contrary to caricatures of wind opposition groups as based in urban elite culture, the population of Searchlight is not particularly wealthy nor does it represent second-home owner communities. The median household income in 2008 was recorded at $24,407. Most of Searchlight's housing consists of double-wide trailer homes.

This historic mining town is now popularly known for its recreational offerings. The Mojave National Preserve, a national park unit famous for its volcanic cinder cones and Joshua trees, lies just west of Searchlight. To the east, the Colorado River winds its way through the desert landscape of the Lake Mead National Recreation Area. Cottonwood Cove, which is 16 miles from Searchlight and within the Lake Mead Recreational area, is one of the West's best largemouth bass fisheries.

The BLM is overseeing the permitting process for the Searchlight Wind project. Searchlight Wind, LLC, a subsidiary of North Carolina based Duke Energy, is proposing to develop a 370 MW facility consisting of up to 140 wind turbines. The BLM reported that the proposed turbine towers would be up to 262 ft tall from the ground to the hub with blades extending up an additional 153 ft. The total height of each turbine would be up to 415 ft (BLM 2009). In comparison, the tallest structure currently in town is the 150 ft flagpole located at a local establishment known as Terrible's Casino. In addition to the turbines, the proposed project will also require the construction of new access roads, a transmission line, two electrical substations, an electrical interconnection facility/switchyard, and a maintenance building.

Project developers purport that the facility will have the capacity to generate enough electricity to power over 100,000 households, assuming an average household use of approximately 9000 kW hours per year. Yet, like other utility-scale electricity projects, this power will not provide "green" energy directly to Searchlight but will instead feed into the transmission lines that travel electricity north to the bright lights of Las Vegas. According to Duke Energy's Managing Director Robert Charlebois, the project will provide local permanent employment for 15 workers (Edwards and Tetreault 2008).

The BLM began the initial stage of the EIS process, known as the "scoping period", in December 2008. During the scoping period, the BLM is required to engage potentially affected federal, state, and local agencies, American Indian tribes and the public in determining issues of concern. The BLM scoping report states that they received 384 unique issue comments regarding the project. One third of these comments expressed concern about wildlife, noise or visual resources

(BLM 2009). The BLM's report noted a range of questions in public comments including:

- Will the placement of wind turbines affect views of scenic areas such as Lake Mohave and the surrounding mountains?
- Will the new facility give Searchlight an industrial look?
- What steps will be taken to minimize visual impacts on the area?
- There are numerous energy projects proposed in the area. How will these be evaluated for past, present, and future cumulative impacts?
- How will quality of life for Searchlight residents change as the wind facility changes the area?

While this project is still early in its permitting stage, with an expected release of the draft EIS in 2011, there has been considerable local rancor about project process and potential landscape impact.

Local conservation groups have been documenting community concerns since the project was first announced in 2008. Basin and Range Watch is one such volunteer-based conservation organization. In addition to the Searchlight case, their website documents a dozen other wind and solar projects in the permitting process in this area. One strategy for influencing local perceptions of the project's impact is the creation of photo simulations. Figure 2 suggests how the project will look from local points of interest. On their website, the caption for the image reads:

Duke Energy's huge turbine array would be too close to this rural town in southern Nevada, say many long-time residents. Here's their photomontage of the proposal, looking eastward from town towards the Colorado River Valley below. Recreationists on Lake Mead would have to look up and see this array from the other side.

Figure 2: Image as it appears on the Basin and Range Watch website[7]

In a piece published after the first BLM town hall meeting, entitled "Town vs Wind Goliath", Basin and Range volunteers write about the local politics of the project. They ask:

> How does a small rural town react when it is chosen to be the centerpiece of an industrial renewable energy project? ... They did not ask for the BLM land around them to be gobbled up for abstract national interests, for projects that will not directly benefit them.

The local politics surrounding the project are complicated by the fact that one of its chief proponents is US Senate Majority Leader, Harry Reid. Reid hails from the small town of Searchlight. A small museum in town and the local elementary school are dedicated to Harry Reid. As a son of a Searchlight miner, Reid has written the only major historical study on this town. His book, *The Camp That Never Failed*, chronicles the cycles of mining boom and bust that eventually lead to Searchlight's current reinvention as a retirement community adjacent to the Lake Mojave Recreational Area. Searchlight was the home of the Quartette gold mine, where a 1903 gold strike made it one of the most productive mining towns in the entire world (Reid 1998:33). After the gold boom, however, the town's population was reduced to just fifty when the new interstate highway bypassed the town. Reid writes that "Searchlight may have not been favored by nature, but in the years after gold was discovered, this desert place developed into a microcosm of a frontier settlement" (1998:5).

Today, Searchlight is a *historic* mining town. Most local residents are unfamiliar with a landscape of active energy extraction. The place politics of Searchlight, particularly its extractive history, contribute to the town's quaintness. The Searchlight Heritage Museum depicts the town's mining and railroad history through photos, artifacts, exhibits, and an outdoor mining park.The arrival of a 350 MW energy project is likely a turn toward an unwanted industrial past. Wind energy is, in fact, relatively foreign to Nevada residents. Searchlight will be the first utility-scale project in the state. The closest utility-scale wind projects are located 250 miles to the southwest in Palm Springs, California.

Given Searchlight's demographics, the project offers little economic benefit to local residents, particularly those who are retired and have no use for the few jobs that will be created by the project. Because the facility is on public lands, this project is also unlikely to provide new tax revenues directly to the town. The BLM will establish a PILT (public land in lieu of taxes) agreement with Duke Energy. The price of the wind lease will be based on the kilowatts generated at the site and will be paid annually to the state of Nevada. The state will determine how much of this revenue gets redistributed back to the county.

How do we understand this one community's opposition to wind development? This case study documents two important characteristics of American wind opposition. First, despite what developers and policy makers may believe, residents of the "New American West" are not necessarily supportive of wind energy development. Shifting demographics and landscape consumption values suggest that utility-scale development may be as contested in the American West as it has been in other regions of the USA. Second, oppositional campaigns are not patiently waiting for bureaucratic permitting processes to unfold. They are organizing and networking early and effectively to challenge not only the technical designs for projects, but their very basic local economic and environmental rationales.

Taken together, these two observations deserve further unpacking. Historically, "Old West" rural communities were heavily dependent on ranching, forestry, farming or mining resources. Yet, rural communities in the West have been experiencing rapid shifts in terms of their economies, social structures and cultural identities (Winkler et al 2007). The rural sociology and geography literatures on the "New American West" have documented that rural in-migrants, including but not limited to "baby boomers", are seeking refuge from "urban problems and suburban sprawl in the small towns, and in rural communities, surrounding public lands, parks, lakes, mountains, and forests" (Jones, Fly and Cordell 2003:223). These new migrants often come with a diverse range of organizational and leadership skills which enable them to challenge long held assumptions about "Old West" values. Rather than create a culture clash with long-time residents, Jones and colleagues argue that "green" migrants elevate existing concerns about protecting and preserving the environment. New in-migrants are particularly interested in protecting the public lands resources that drew them toward "seeking natural beautiful settings" (Winkler et al 2007:481).

Searchlight fits the characteristic of an amenity-based community within short driving distance of a national park or national monument. Searchlight is, however, unlike the typical amenity-based New West town in at least two ways. Winkler et al estimated that the median income of more than half of the "New West" towns was $79,500 in 2000. In contrast, the median income in Searchlight was $24,407 in 2000. Searchlight is also not represented by the stereotypical New West seasonal resident with an appetite for "lattes, turquoise jewelry, antler art and fine dining" (Winkler et al 2007:497). Wind development is likely to be even more contentious around more typical "New West" towns with higher incomes and a larger percentage of seasonal residents.

Nearly a decade ago, geographer Michael Pasqualetti suggested that communities would react strongly to the aesthetics of wind turbines because we expect "permanence" in our landscapes (2000). This perception is "rudely violated" when abrupt and fundamental landscape change occurs in a way that threatens a sense of place. This is especially true in the western USA, where wind power developments challenge the lingering image of "Big Sky Country". Pasqualetti wrote that "Open space remains the West's greatest attribute and attraction", and is understood as an "inalienable right of all those with the luck to have been born there or—as some believe—the sense to have moved there" (2000:390). The pace and scale of wind development in the West, combined with shifting demographics and place politics, makes new energy development particularly contentious.

Since the draft EIS statement is yet to be released, it is impossible to know the fate of this project. Given these realities, the Searchlight case can be understood as a cautionary tale for wind energy developers who have assumed their brand of green power is welcome in the American West. While "aspirational ruralism" may be playing a role in wind opposition movements, it would be misleading to equate this with straight-up NIMBYism. Opposition to the project is based on many issues, including scale, emerging landscape values and potential local economic benefits. Nevada is not unique in this regard. Neighboring states like Montana and Wyoming are similarly caught in a wind "gold rush". For example, in a recent interview with the Associated Press, the Governor Freudenthal of Wyoming, a state known for oil and natural gas exploration, stated:

> I appreciate the fact that people can say [wind] has great environmental benefits, but that's people who don't live next to them, or whose wildlife habitat isn't being disrupted, or the bird population isn't being affected, or whose view isn't being altered.

He went on to add that wind companies "are not entitled to a free ride" (Associated Press 2010).

Moving Forward

The previous sections described the shifting terrain of wind opposition in America. Drawing on Leo Marx's evocation of the inherent contradictions and possibilities embodied by the "middle landscape" ideal, this final section asks if wind energy's impacts on rural landscapes can be reconciled with renewable energy targets. In other words, can expansive wind energy development have a place in the rural landscape imaginary that remains so important to American cultural constructions of nature? While the critical geography literature has called attention to the social and political consequences of an aestheticized spatiality, wind opposition campaigns point toward a set of new normative tensions in the aligning of a new energy economy. Rather than sideline these concerns as examples of NIMBYism, I argue that it is important to struggle with questions such as: Who decides which landscapes and communities bear the burden of new energy development in the national and international interest? What role should residents, old and new, play in determining landscape–technology compatibility, scales of design, pace of development and acceptable mitigation?

As Leo Marx argued nearly 50 years ago, American pastoralism embodies a powerful contradiction toward nature by celebrating rural landscapes as "wild" yet embracing industry and commercialism as a means of sustaining this utopia. American pastoralist commitments continue to have a resilient ability to both thwart and embrace new symbols of progress that challenge the viability of the "middle landscape" ideal. In his later works, Marx came to argue that American pastoralism, particularly that which became manifested in strands of environmentalism, had the potential to offer an alternative path for our industrialized of society (Cannavò 2001). In the Epilogue to the 1964 volume, Marx claimed that modern society needed to create "new symbols of possibility" that responded to the machine's sudden and profound entrance into the garden (1964:365).[6]

In *Working Landscapes*, Peter Cannavò argues that polarized social movements that either defend preservation or defend development fail to recognize the value of a "working landscape" in strengthening the ability of local residents to create planning approaches that reconcile place attachment with economic development (2007). Cannavò uses bioregionalism as a model for addressing how land use policies can address the "crisis of place". He advocates for deliberative planning processes that move beyond administrative rationalism toward flexible and collaborative conservation. With the Searchlight case still in flux, the BLM and local conservation advocates have an opportunity to pursue a new posture toward wind development where local residents negotiate project characteristics in ways that respect local landscape values.

Land use conflicts have dominated the political culture of the American West since the founding of the BLM nearly 60 years ago (McCarthy 2002). Conflict resolution organizations, conversation groups and planning think tanks have an important role to play in balancing regional climate change actions with quality of life concerns. As Pasqualetti argued a decade ago:

> If wind energy is to expand, so too will wind-energy landscapes and the attention paid to them by the public. If developers are to cultivate the promise of wind power, they should not intrude on favored (or even conspicuous) landscapes, regardless of the technical temptations these spots may offer (2000:392).

Following this prescription, there are scenic areas of the West that should be off limits to renewable energy development. The Natural Resources Defense Council has argued as much through its "Clean Energy in the Western US" Google Earth interactive mapping tool.[7] This resource helps environmentalists, renewable energy developers, and utility companies understand where new installations are likely to meet great public opposition. The NRDC website argues that because the West is home to remarkable wildlands, diverse wildlife and irreplaceable cultural resources, it is "vital to find the best sites for new clean energy projects and transmission lines, so that America can harness renewable power while doing the least damage to the Western environment". Oriented toward developers, the NRDC map argues that knowing the location of ecological and culturally sensitive lands is the first step toward ensuring that new projects and new transmission lines are built in the public interest.

Cultivating a new wind development ethic also must more fundamentally involve land use planners in energy siting decisions. The Lincoln Institute's 2010 study "Planning for climate change in the American West" described the lack of public support for climate change mitigation and the need for local action and citizen participation. The report argues that:

> It is important to recognize that there is an inherent challenge to reaping the economic benefits of climate-related policies. While local communities may bear the brunt of the costs for policies and actions to address climate change, the benefits created can often extend beyond their communities, which can also contribute to slow adoption of mitigation and adaptation strategies (Carter and Culp 2010:35).

If wind energy development is seen as part of long-range climate solutions, it must be recognized that rural communities cannot be asked to bear the burdens of carbon mitigation alone. The report's authors suggest that state climate action plans, which can include aggressive wind development, need to produce region-wide strategies that involve urban and rural areas alike. The wind opposition discourse is charged with rural–urban conflicts over climate change mitigation and new energy development. This has been evident in the Searchlight case where local residents argue that power generated by the new wind farm will feed Las Vegas' over-the-top conspicuous energy consumption at the expense of the rural character they love.

Conclusion

Concerns about climate change, energy security, and economic instability have produced a resounding public call for wind energy development in America. Yet, can "Big Wind" succeed in the American West? There is a considerable "social gap" between broad national support for wind energy and vocal local resistance (Bell, Gray and Haggett 2005). A growing social movement is calling attention to the rural landscape impacts associated with these new energy developments, particularly in terms of wildlife, sound and visual impacts felt by those communities living near wind projects. Through active campaigns to "preserve" rural landscapes from "industrial" wind energy development, wind opposition organizations are contesting who speaks for the productive and consumptive qualities of rural landscapes.

While it is inevitable that rural energy geographies must shift to constitute a new low-carbon energy economy, it is also necessary that wind energy developers, rural citizens and government officials negotiate an important tension in the American pastoral imaginary vis-à-vis green industrialization. Social movement organizations have an important part to play in forging a "middle landscape" sensibility that dynamically values socio-cultural resources without falling prey to a form of "sentimental pastoralism" that fails to acknowledge our dependency on technology and industry. Rather than fall back on accusations of NIMBYism on the one hand, or conspiracy and greed on the other, a new course for wind development must involve regional models of collaborative development where urban and rural residents are asked to respectfully and constructively grapple with crucial issues of scale, design, equity and democracy.

Acknowledgments

I am grateful for the constructive comments made by the editor and the three anonymous referees. Special thanks to Max Boykoff, Pete Newell and Emily Boyd, as well as the other panelists and audience members of the 2009 AAG sessions on the New Carbon Economy, for guidance. Finally, I am grateful to Birgit Muehlenhaus for her mapping assistance. This research has been supported by the National Science Foundation (SES# 0724672).

Endnotes

[1] After publication of this article, Duke has scaled back its original project. As of 2011, the proposed project is set at 87 turbines, generating 200 MW at a cost of approximately US \$400 million.

[2] These case studies and maps are accessible at http://www.macalester.edu/windvisual/

[3] Personal interview with Eric Rosenbloom, President, National Wind Watch, 11 March 2009.

[4] The "green jobs" estimate ranges widely. I have relied on NREL numbers because they are the industry standard. The actual jobs generated per project also depend on the turbine models chosen and their maintenance requirements.

[5] The term "Starbucks effect" is used colloquially in wind industry circles. The term was directly referenced in a presentation by Trey Cox titled "Lessons from the siting

wars" at the Windpower 2009 Conference and Exhibition in Chicago, Wednesday 6 May 2009.

[6] Marx's scholarship has met considerable critique for its broad claimsmaking and lack of concrete examples of particular technological impacts on "the garden". See Meikle's 2003 essay, "Leo Marx's *The Machine in the Garden*", for an exploration of the challenges leveled against Marx by American Studies and History of Technology scholars. Cannavò's 2001 essay critiques Marx's lack of attention to alternative variants on sentimental pastoralism offered by the nascent environmental movement.

[7] This tool is available at http://www.nrdc.org/land/sitingrenewables/default.asp

[8] These images are printed with the permission of Basin and Range Watch. They are posted at http://www.basinandrangewatch.org/SearchlightUpdates.html

[9] Adapted from Table A presented in J. B. Graham et al. (2009) "Public perceptions of wind energy developments: Case studies from New Zealand", Energy Policy 37(9): 3349–3357.

References

American Wind Energy Association (2010) *Year End 2009 Market Report*. Washington, DC: AWEA

American Wind Energy Association and Canadian Wind Energy Association (2009) *Wind Turbine Sound and Health Effects: An Expert Panel Review*. Washington, DC: AWEA

Asmus P (2000) *Reaping the Wind: How Mechanical Wizards, Visionaries, and Profiteers Helped Shape Our Energy Future*. Boston: Island Press

Associated Press (2010) Wyoming's Governor Dave Freudenthal: Wind companies are not entitled to a free ride. 12 March

Barry J, Ellis G and Robinson C (2008) Cool rationalities and hot air. *Global Environmental Politics* 8(2):67–98

Bell D, Gray T and Haggett C (2005) The social gap in wind farm siting decisions. *Environmental Politics* 14(4):460–477

BLM (2009) Scoping summary report for the Searchlight Wind Energy Project. Bureau of Land Management. http://www.blm.gov/nv/st/en/fo/lvfo/blm_programs/energy.html (last accessed 30 November 2009)

Cannavò P (2001) American contradictions and pastoral visions: An appraisal of Leo Marx's *The Machine in the Garden*. *Organization & Environment* 14:74–92

Cannavò P (2007) *Working Landscapes: Founding, Preservation, and the Politics of Place*. Cambridge: MIT Press

Cappiello D and Hebert H J (2009) Salazar pushes for wind energy. *Associated Press* 9 March

Carter R and Culp S (2010) *Planning for Climate Change in the West*. Cambridge, MA: Lincoln Institute of Land Policy

Cass N and Walker G (2009) Emotion and rationality: The characterisation and evaluation of opposition to renewable energy projects. *Emotion, Space and Society* 2:62–69.

della Porta D and Diani M (1999) *Social Movements: An Introduction*. Oxford: Blackwell

Department of Energy (2008) *20% Wind Energy by 2030*. Oakridge, TN: Office of Scientific and Technical Information

DePillis P (2007) Turbine foes try to forge national opposition movement. *Greenwire* 24 September

Devine-Wright P (2005) Beyond NIMBYism. *Wind Energy* 8:125–139

Duncan J S and Duncan N G (2004) *Landscapes of Privilege: The Politics of the Aesthetic in an American Suburb.* New York: Routledge

Edwards J and Tetreault S (2008) Wind farm floated in state, Duke Energy proposes project near Searchlight. *Las Vegas Review* 19 December

Escobar A (2001) Culture sits in places. *Political Geography* 20:139–174

Gipe P (2009) *Wind Energy Basics.* White River Junction, VT: Chelsea Green Press

Hindmarsh R and Matthews C (2008) Deliberative speak at the turbine face. *Journal of Environmental Policy and Planning* 10(3):217–232

Jones R, Fly M and Cordell K (2003) Green migration into rural America: The new frontier of environmentalism? *Society and Natural Resources* 16:221–238

Kahn R (2000) Siting struggles. *The Electricity Journal* 13(2):21–33

Kempton W, Firestone J, Lilley J, Rouleau T and Whitaker P (2005) The offshore wind power debate: Views from Cape Cod. *Coastal Management* 33:119–149

Marx L (1964) *The Machine in the Garden: Technology and the Pastoral Ideal in America.* New York: Oxford University Press

McCarthy J (2002) First world political ecology: Lessons from the Wise Use movement. *Environment and Planning A* 34(7):1281–1302

Meikle J L (2003) Leo Marx's *The Machine in the Garden. Technology as Culture* 44:147–159

Pasqualetti M (2000) Morality, space, and the power of wind-energy landscapes. *Geographical Review* 90(3):381–394

Pasqualetti M, Gipe P and Righter R (2002) *Wind Power in View: Energy Landscapes in a Crowded World.* San Diego: Academic Press

Phadke R (2010) Steel forests and smoke stacks: The politics of visualisation in the Cape Wind controversy. *Environmental Politics* 19(1):1–20

Reid H (1998) *Searchlight.* Las Vegas: University of Nevada Press

Robbins P, Meehan K, Gosnell H and Gilbertz S (2009) Writing the new West. *Rural Sociology* 74(3):356–382

Schroeder R A and Albert K E (2006) Political ecology in North America: Discovering the Third World within? *Geoforum* 37(2):163–168

Smith B (2007) *Wind Energy: New Challenges, New Opportunities.* Golden, CO: National Renewable Energy Lab. http://www.cresenergy.org/documents/meetings/CRES_Meeting_Wind_25Oct07_bsmith.pdf (last accessed 3 December 2009)

Sower J (2006) Fields of opportunities. *Great Plains Quarterly* 26(2):99–112

Szarka J (2004) Wind power, discourse coalitions and climate change. *European Environment* 14:317–330

Tuan Y (1974) *Topophilia: Study of Environmental Perception, Attitudes and Values.* Englewood Cliffs, NJ: Prentice Hall

Van Der Horst D (2007) NIMBY or not? *Energy Policy* 35(5):2705–2714

Walker P and Fortmann L (2003) Whose landscape? *Cultural Geographies* 10:496–491

Winkler R, Field D, Luloff A E, Krannich R S and Williams T (2007) Social landscapes of the inter-mountain west: A comparison of "old West" and "new West" communities. *Rural Sociology* 72(3): 478–501

Wolsink M (2007) Planning of renewables schemes. *Energy Policy* 35:2692–2704

Woods M (2003) Deconstructing rural protest: The emergence of a new social movement. *Journal of Rural Studies* 19:309–325

Index

Note: page numbers in *italics* refer to Figures; those in **bold** to Tables; notes are indicated by suffix 'n' (e.g. "61n[15]" = page 61, note 15)